LAYERED MATERIALS

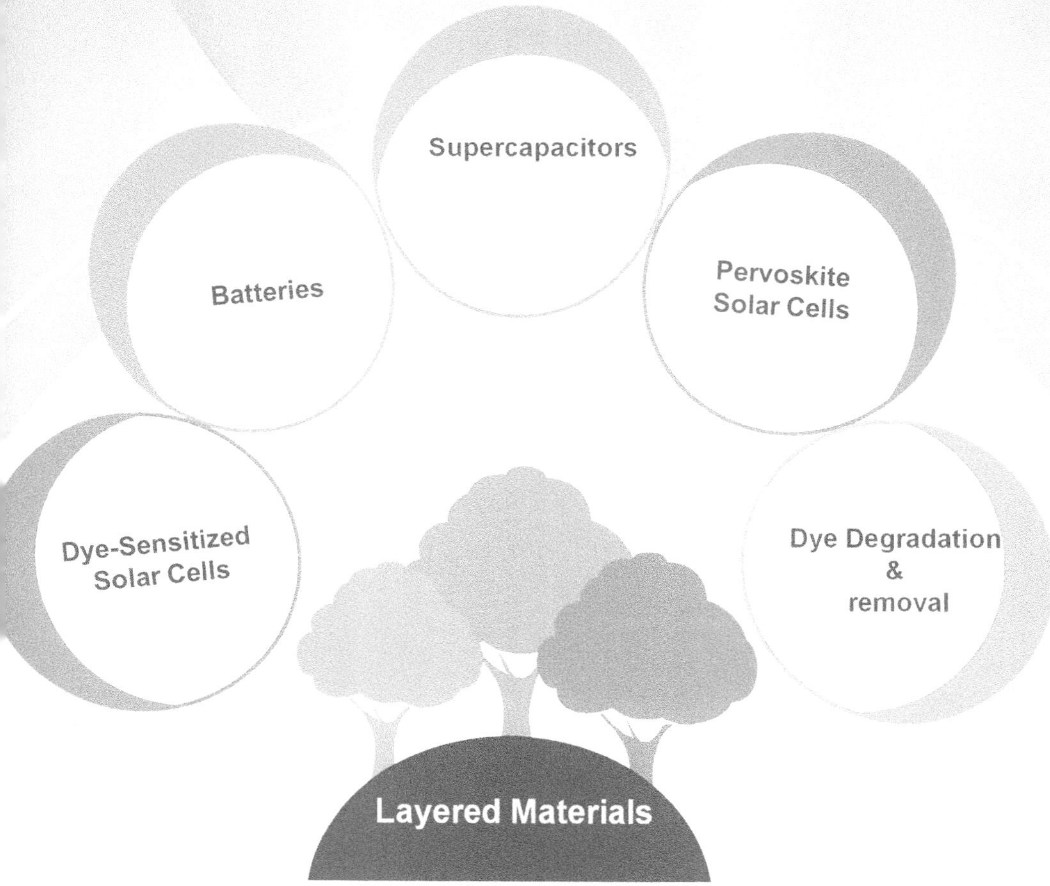

LAYERED MATERIALS

Synthesis and Applications in Energy and Environmental Remediation

edited by

A. Pandikumar

K. Jothivenkatachalam

S. Moscow

JENNY STANFORD PUBLISHING

Published by

Jenny Stanford Publishing Pte. Ltd.
101 Thomson Road
#06-01, United Square
Singapore 307591

Email: editorial@jennystanford.com
Web: www.jennystanford.com

British Library Cataloguing-in-Publication Data
A catalogue record for this book is available from the British Library.

Layered Materials: Synthesis and Applications in Energy and Environmental Remediation

ISBN 978-981-5129-17-5 (Hardcover)
ISBN 978-1-003-50831-1 (eBook)

Contents

Preface

Layered materials have been the central tool of the photocatalytic material research community owing to their unique electronic, optical, physical, chemical, and mechanical properties in layered form compared with their bulk form. They have been utilized in electronics, catalysis, energy, environment, and storage device applications. This book offers the state of the art of research on the synthesis, properties, characterization, and electrochemical applications of layered materials and highlights the present challenges related to environmental issues. It provides fundamental, experimental, and theoretical knowledge about layered materials, including graphene, graphitic carbon nitride, boron nitride, and MoS_2-, Li-, and Mn-rich oxide materials and MXenes, and elaborates their synthesis, physicochemical properties, and structure-property-electrochemical performance. The book also covers the recent progress in developing layered material-based electrochemical energy conversion systems and storage devices and identifies immediate research needs and directions in developing layered material-based systems for future applications. It emphasizes how the activity of layered materials can be improved by modifying their structural, electrical, and optical properties for the fabrication of storage devices. The book is a useful reference for academics, engineers, and industrial researchers and will help early-career researchers in establishing their research skills in the area of layered materials.

Dr A. Pandikumar
Senior Scientist
Electro Organic and Materials Electrochemistry Division
CSIR-Central Electrochemical Research Institute
Karaikudi 630003, Tamil Nadu, India

Dr K. Jothivenkatachalam
Professor and Head
Materials Research Laboratory for Energy and Environmental Applications
University College of Engineering—BIT campus
Anna University, Tiruchirappalli 620024, Tamil Nadu, India

Dr S. Moscow
Assistant Professor
Materials Research Laboratory for Energy and Environmental Applications
University College of Engineering—BIT campus
Anna University, Tiruchirappalli 620024, Tamil Nadu, India

Chapter 1

Graphene-Based Dye-Sensitized and Perovskite Solar Cells

Ramachandran Krishnapriya,[a] Devika Laishram,[a] Elayappan Vijayakumar,[b] Sudhi Mahadevan,[c] Rakesh Kumar Sharma,[a] and Hai-Gun Lee[b]

[a]*Sustainable Materials and Catalysis Research Laboratory, Department of Chemistry, Indian Institute of Technology Jodhpur, Jodhpur, Rajasthan 342037, India*
[b]*Department of Materials Science and Engineering, Korea University-Seoul Campus, 145 Anam-ro, Seongbuk-gu, Seoul 02841, Republic of South Korea*
[c]*Department of Materials Science and Engineering, City University of Hongkong, 83 Tat Chee Avenue, Kowloon Tong, Hong Kong*

krishnapriyachem@gmail.com, evijayakumarindia@gmail.com

Environmentally amenable renewable energy sources, such as biomass, solar, hydroelectric, wind, and hydrogen, are indispensable for the existence and development of social evolution. Amongst these sources, sunlight is plentiful and inexhaustible, and has gained considerable significance in recent energy research. This clean and sustainable form of energy can be effectively converted to electricity via the means of solar cells. Notably, the third-generation (3G) solar cells, such as dye-sensitized solar cells (DSSCs) and perovskite solar cells (PSCs), find specific attention

Layered Materials: Synthesis and Applications in Energy and Environmental Remediation
Edited by Alagarsamy Pandikumar, Kandasamy Jothivenkatachalam, and Subramanian Moscow

ISBN 978-981-5129-17-5 (Hardcover), 978-1-003-50831-1 (eBook)
www.jennystanford.com

amongst the most sustainable technologies, due to their low cost with less carbon footprint. However, the impediment is to use materials that are inexpensive, ecological, and possess attractive properties for catalytic applications. In this regard, graphene, a one-atom thick, monolayer of graphite with a unique two-dimensional (2D) structure, is highly alluring. It has attracted significant research interest, owing to its exceptional electronic, thermal, mechanical, and charge transport properties. These remarkable features of graphene, along with the intrinsic characteristics of the carbon, make this honeycomb-structured material a promising candidate for various energy conversion and storage device applications. Over the past few years, significant effort has been made in the design and fabrication of graphene-based electrode materials for their extensive use in each part of solar cells (anode, cathode, electrolyte, and transparent conducting oxide layer) and has attained prodigious performance. This chapter summarizes the state-of-the-art research of different graphene-based materials that can play a diverse role as an active layer, sensitizer, counter-electrode, hole transport layer (HTL), and electron transport layer (ETL) in DSSC and PSC device applications. In this chapter, the synthesis, morphological features, properties, and performance of graphene and its related compounds, followed by the crucial issues, challenges, and future perspectives are addressed.

1.1 Introduction

Overconsumption of natural resources and environmental pollution have intensified climate change and outpaced the sustainable capacity of our ecosystem. Consequently, developments of sustainable and clean energy technologies, which are more efficient and cost-effective, are getting enormous scientific attention. In line with this precedence, the exploration of renewable energy technologies based on biomass, wind, wave, geothermal, and solar energies is essential to meet the current energy consumption. Amidst the various renewable energy sources, solar energy is the leading viable solution to meet existing worldwide energy demand, as it is freely available and clean [1–3]. The devices that can harness and convert solar energy directly to useful electrical energy are called solar cells and work through the photoelectric effect. Over many years, the first-generation solar cells (based on crystalline silicon) and the second-generation solar cells (based on amorphous silicon, CdTe, and CIGS) were used widely. However, these devices suffer from the high cost of manufacturing, installation, and environmental unfriendliness, which confine their extensive commercial applications. As a 3G solar cell, DSSCs and PSCs are recognized as potential candidates to meet the ever-growing clean energy demand, owing to their simple solution-processable synthesis, cost-effective fabrication process, and environmental friendliness [4, 5]. Hence, they can provide excellent large-scale solar electricity generation at an economically viable cost.

1.2 Dye-Sensitized Solar Cells (DSSCs)

A DSSC is a modern electrochemical energy conversion device that was inspired from nature by the breakthrough work of O'Regan and Gratzel in the laboratory at the Ecole Polytechnique Fédérale de Lausanne in 1991 [6]. The generic DSSC device comprises four components:

(i) **Photoanode:** It consists of a nanostructured wide bandgap n-type semiconductor (TiO_2, ZnO, SnO_2, etc.) finely coated on a transparent conductive fluorine-doped tin oxide (FTO) glass substrate.

(ii) **Sensitizer:** It consists of a ruthenium metal complex with carboxylated bipyridine ligand-based inorganic dye molecules to absorb sunlight.

(iii) **Redox electrolyte:** It is an I^-/I^{3-} based liquid electrolyte to facilitate electron transfer.

(iv) **Counter electrode (CE):** It is generally made of platinum electrocatalyst for the transfer of electrons from the external load to the counter electrode [7–8].

A schematic illustration of a typical DSSC device and its working principle is given in Fig. 1.1.

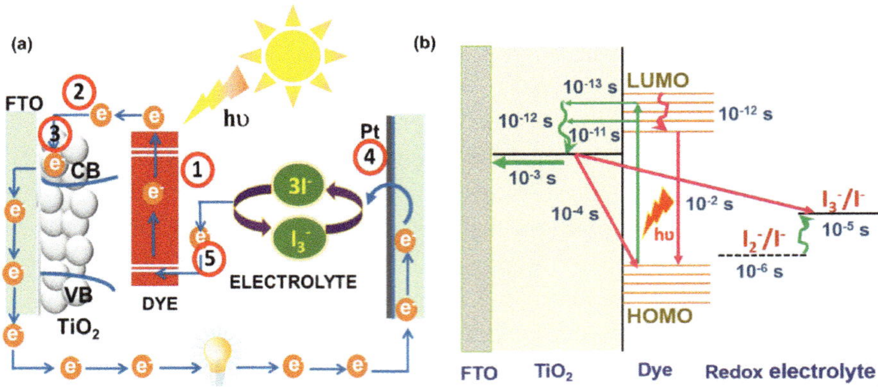

Figure 1.1 (a) Schematic representation of different components and the electron transfer mechanism in a DSSC. (b) Typical time constants occur in the DSSC under illumination.

There are five different steps included in the working process of a DSSC device. Initially, the incident light energy (photon) reaches the anode and is absorbed by a photosensitizer. Afterward, the promotion of electrons from the ground state to the excited state of the photosensitizer occurs. This process is called *photoexcitation,* and the excited electrons are highly unstable with a lifetime of several nanoseconds. So, in the second step, these electrons are introduced into the conduction band of the semiconductor (TiO_2) electrode, which lies lower to the excited state of the sensitizer, which is called the *electron injection* step. Here, the TiO_2 photoanode absorbs photons of the UV region, and subsequently, the dye molecule gets oxidized.

Further, the injected electrons in the conduction band of the semiconductor are transported and diffused toward the FTO and reach the CE over the external circuit by the *electron diffusion* process. In the fourth step, the oxidized redox mediator (I^{3-}), diffuses toward the CE, and there it is reduced to I^- ions (*reduction of redox electrolyte*). The final step is the *dye-degeneration* in which the oxidized dye receives electrons from the I^- ion of the redox mediator, resulting in the regeneration of the ground state and subsequent oxidation of I^- to I^{3-}. Accordingly, the DSSC device generates electricity from sunlight deprived of any permanent chemical transformation [9, 10].

In recent years, DSSCs have realized a photoelectric conversion efficiency (PCE) above 13% at the laboratory scale and have advanced quickly, owing to their substantial benefits, including cost-effectiveness, simple fabrication process, and performance under diffused light [11, 12]. However, DSSCs still suffer from some disadvantages, such as the incapability to absorb full absorption of solar energy and the photobleaching of dyes. These demerits have stimulated solar energy researchers to develop an alternative solid dye material.

1.3 Perovskite Solar Cells (PSCs)

A PSC is a special type of 3G solar cell device that uses perovskite structured material as an active light-harvesting layer. The crystal structure of perovskites is similar to $CaTiO_3$, with a general chemical formula ABX_3, where A is a large cation, and B a smaller cation, which has octahedral coordination by six X anions as shown in Fig. 1.2 (a). These materials have engrossed extensive consideration, due to the cubic lattice-nested octahedral (Oh) layered structures and the inimitable photoelectric properties such as less exciton binding energy, large optical absorption coefficient, thermal, electrical, and magnetic characteristics. The perovskite materials employed in PSCs are organic-inorganic metal halides. In such compounds, the cation A (methylammonium, CH_3, MA^+, or formamidinium, FA^+) is located in the vertex of the face-centered cubic lattice, and the metal cation B (Pb^{2+}, Sn^{2+}, etc.) and halogen anion X (Cl^-, Br^-, I^-, or coexistence of several halogens) fill up the core and apex of the octahedra, correspondingly [13]. In a typical lead halide perovskite, A is a large inorganic cation such as Cs^+, or organic cations such as methylammonium or formamidinium, B is Pb^{2+}, and X is a halide anion (I, Br, Cl). The metal-halogen octahedra are joined to make a stable three-dimensional (3D) network structure [13]. Commonly used perovskite materials in PSCs are hybrid organic-inorganic Pb or Sn halides (methylammonium lead halides), and all-inorganic cesium leads halides. PSCs stemmed from DSSCs in 2009 by Miyasaka and co-workers, who employed organolead halide perovskite NCs such as $CH_3NH_3PbBr_3$ to efficiently sensitize the semiconductor TiO_2 for light conversion in DSSC devices [14]. These 3G devices emerged as an innovative family of solar cells during the past decade and are recognized as a prospective candidate for commercialization, owing to their cost-effectiveness, scalability, and extraordinary power conversion efficiency (PCE) of about 3% in 2009 to over 22.1% in 2016 and 24% in 2019 (for a Pb-halide PSC with area 0.1 cm^2) [14–16].

The components of PSCs are similar to DSSCs, which consist of semiconducting TiO_2 as an ETL, a layer of perovskite materials for light-harvesting, an HTL, and an electrode, as shown in Fig. 1.2(b). The power conversion efficiency of these devices is mostly controlled by factors like short circuit current (I_{sc}), open-circuit voltage (V_{oc}), and fill factor (FF). The important processes that take place include (a) light absorption, (b) charge generation, (c) charge transport, and (d) charge extraction [17, 18]. When sunlight falls on the perovskite layer, it absorbs photons to yield excitons (electron-hole pairs), owing to the difference in the exciton binding energy of the perovskite materials, these can form free carriers (free electrons and holes) to produce current or can recombine into excitons [13]. Here, a thinner film is used for light absorption, which provides shorter paths for charge transport, which, in turn, helps to avoid the possibility of recombination, thus reducing the energetic losses and increasing the final voltage of the device. Further, these free electrons and holes are collected by an electron transport material (ETM) and a hole transport material (HTM). Electrons are transferred from the perovskite material to the semiconductor oxide layer (TiO_2), and finally collected by FTO. Simultaneously, the holes are transferred to the HTM layer and are collected by the metal electrode (Ag). Lastly, the FTO and metal electrode are connected, and the photocurrent is produced in the outer circuit [13]. The high electron mobility and conductivity of intermediate layers help electrons or holes reach the electrode. Generally, metal oxides like TiO_2 and ZnO, semiconducting polymers like PEDOT:PSS and P3HT, fullerene acceptors like PCBM, etc. have been successfully used so far as electron or hole transport layer [19–21].

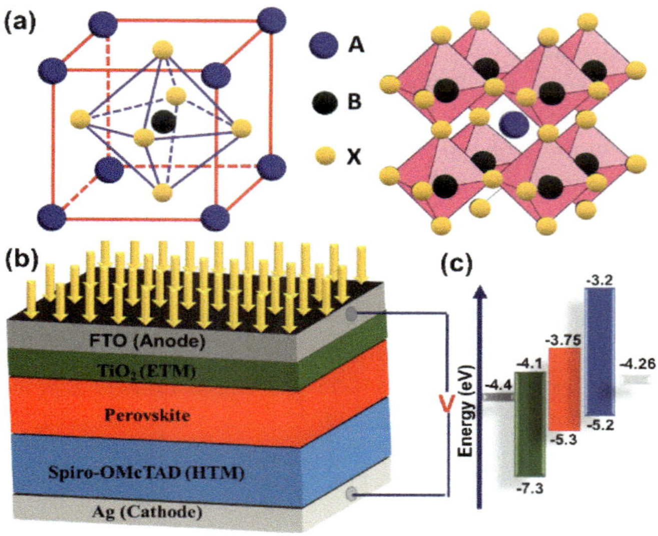

Figure 1.2 (a) Crystal structure of perovskite material. (b) Device structure of a PSC. (c) Energy diagram of individual layers of PSC device.

1.4 Graphene: The Wonder Material

Graphene is a nanocarbon material consisting of an atomic layer of sp^2 hybridized carbon arranged in a honeycomb lattice with a 2D sheet structure. This wonderful material, with a plethora of applications, was isolated in 2004 by Novoselov et al. at Manchester University, UK [22]. The unique advantages of this material such as a large theoretical specific surface area (2630 $m^2\ g^{-1}$), high intrinsic charge carrier mobility (200 000 $cm^2\ V^{-1}\ s^{-1}$), high thermal conductivity (~5000 $W\ m^{-1}\ K^{-1}$), high optical transmittance (~97.7%), high Young's modulus (~1.0 TPa), and good electrical conductivity have come from quantum effects of the single atomic layer. Due to these exceptional properties, it finds an excess of unique properties, such as being the thinnest and the strongest material ever measured. The outstanding electrical, thermal, mechanical, electronic, and optical properties of graphene have fascinated scientists both from fundamental and applied research fields [23]. It exhibits high specific surface area, chemical stability, optical transmittance, elasticity, porosity, biocompatibility, and band gap tuning. Moreover, these properties of graphene can be carefully tuned by simple chemical functionalization [24]. All these properties make graphene an appropriate candidate for transparent flexible conducting electrode application in various optoelectronic devices. Over the past few years, graphene has proved to be a potential alternative for commonly used transparent metal oxide (TCO), particularly indium tin oxide (ITO), and fluorine-doped tin oxide (FTO). The main benefit of employing graphene in solar cells is to create a perfect window for bringing wide ranges of photon energy. Also, graphene can offer higher charge transfer (CT) kinetics at the interfaces and could also fabricate flexible as well as robust architecture to provide superior heat dissipation [25]. Structurally, graphene is considered to be the fundamental building block for all graphitic materials. It can be wrapped up into zero-dimensional (0D) fullerenes, rolled into one-dimensional (1D) nanotubes, and stacked into three-dimensional (3D) graphite, as shown in Fig. 1.3 [26].

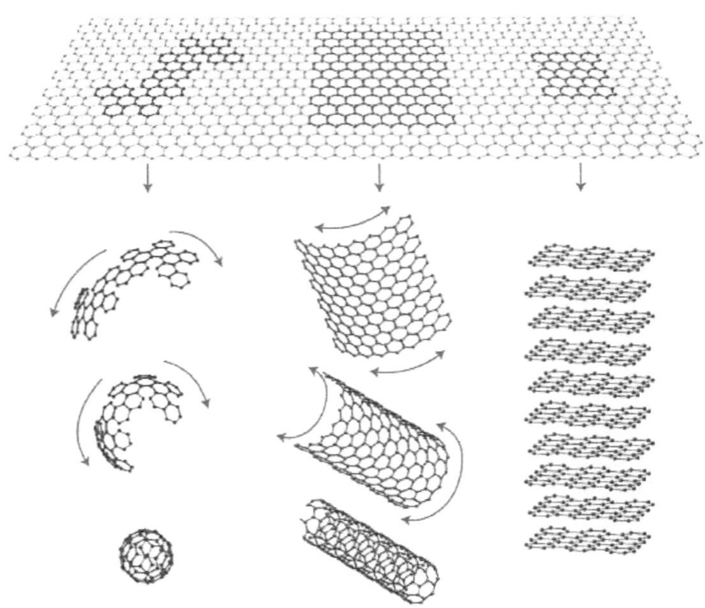

Figure 1.3 Transformation of graphene into buckyballs, 1D nanotubes, and 3D graphite [26].

1.4.1 Synthesis of Graphene

Graphene can be synthesized mainly by two approaches: top-down and bottom-up. In the top-down approach, stacked layers of graphite are broken apart to get the single graphene sheets, while in the bottom-up method, graphene is synthesized from different carbon-containing sources. The first method used to isolate graphene was micromechanical cleavage or 'scotch tape' or 'peel-off' method, which includes the exfoliation of graphite using an adhesive tape to cleave the layers apart, and this yields the highest quality of graphene [22]. But for large-scale production, mechanical cleaving (exfoliation), chemical exfoliation, chemical synthesis, and thermal chemical vapor deposition (CVD) are the frequently used approaches. Apart from these, unzipping nanotube and microwave synthesis methods are also used. Chemical methods are the most recognized approach to obtaining bulk-scale graphene materials. The exfoliation and reduction of graphite oxide have received the most attention among other preparation methods. In this method, graphene oxide (GO) has been synthesized by the oxidation of graphite using concentrated acids and strong oxidants, and intercalating compounds (e.g., H_2SO_4, HNO_3, $KMnO_4$, $KClO_3$, $NaClO_2$). Among these, the 'Hummers method' remains the most widely used process [27]. In this process, graphite is oxidized to GO with a water-free mixture of concentrated sulfuric acid, sodium nitrate, and potassium permanganate. During this procedure, the interlayer spacing of graphite increases, which enables the fine dispersion of graphene in appropriate solvents. Finally, the dispersed GO is converted to graphene by dimethylhydrazine reduction at 80 °C for an extended time. Currently, the modified Hummers method is used for the preparation of large-scale production of graphene layers [28–30]. Figure 1.4 describes the major synthesis approaches of graphene.

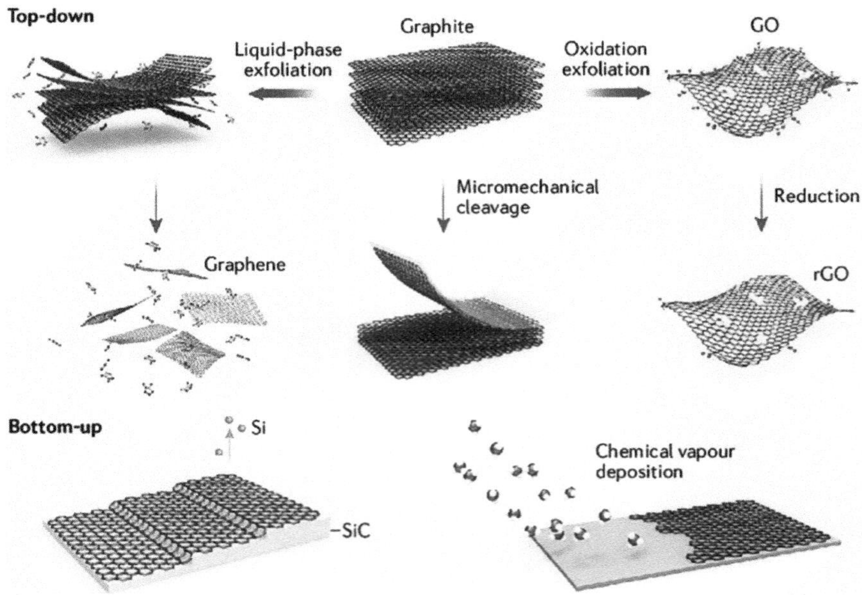

Figure 1.4 Major reported top-down and bottom-up synthesis approaches of graphene such as micromechanical cleavage of graphite, oxidation exfoliation, liquid-phase exfoliation, and chemical vapor deposition [31].

1.4.2 Graphene in DSSCs

1.4.2.1 As a photoanode

DSSC photoelectrode consists of a sensitized semiconductor oxide (TiO_2, ZnO, SnO_2, etc.) layer coated on a transparent conducting glass substrate (FTO/ITO). In order to make the proper working device, the photoanode should be of high surface area for maximum dye-loading, an improved interfacial area between dye and electrolyte, and improved conductivity to avoid recombination. Graphene can be successfully applied mainly as a transparent conducting electrode and also as an electron promoter in the semiconductor oxide layer.

1.4.2.2 As a transparent conducting electrode

The TCO of DSSC is usually made of ITO, which is highly brittle and unstable at high temperatures. Apparently, materials with flexibility, mechanical stability, chemical stability, and catalytic activity are highly desired. The alternative electrode often used is FTO, owing to its stability toward harsh chemical and thermal treatments (up to ~700 °C) and low cost. However, the FTO still suffers from high resistivity of 15 Ω/sq [32]. Distinctive properties of graphene, such as high transparency, good thermal stability, room temperature carrier mobility, and low impedance level, have engrossed excessive research interest in using graphene as a photoanode material in DSSC [33]. The first report on the application of graphene as a transparent conducting layer was done by Wang et al. in 2008 [34]. These graphene films are synthesized from exfoliated GO, followed by thermal reduction. Further, Bi et al. applied the growth of CVD-derived graphene on silica-based glass substrates and demonstrated it as a current collector. However, low efficiency was obtained due to high resistance compared to FTO [35]. Advanced studies on TCO films based on graphene are rapidly progressing.

1.4.2.3 As a semiconducting layer

Recombination loss in the DSSC hampers its efficiency. Many studies on improving the semiconductor oxide layer using graphene were reported. Various forms of graphene, such as an ultra-thin layer, nanofibers, and quantum dots, were used in DSSCs. As of now, various metal oxides, such as TiO_2, ZnO, SnO_2, and PbO, have been mixed with graphene to form efficient nanocomposite materials for PV applications. The first report on the incorporation of graphene in the semiconducting layer was described by Kim et al. in 2009 [36]. A nanocomposite of TiO_2 and reduced graphene were employed as an interfacial layer between FTO and nanocrystalline titania (NC-TiO_2), which lowers the surface roughness at the interfacial layer and is also capable of delivering better adhesion of NC-TiO_2 to FTO. An increase in PCE from 4.89% to 5.26% with a 54-mV increment in the V_{oc}, due to the hindrance of the back-transport reaction in the FTO/TiO_2 interface was observed. There are many factors that affect the efficiency like the ratio of the incorporated graphene and the host semiconductor oxide, the particle size of the monocrystalline semiconductor, the size and thickness of the graphene layer, etc. [37]. Another remarkable work of graphene TiO_2 composite photoanode was reported by Chen et al. [38]. By using spin-coated thin layers of graphene as a blocking layer, a PCE improvement from 5.80% to 8.13% was achieved in comparison to other reported graphene/TiO_2-based DSSCs [39, 40]. To improve the charge collection efficiency, either an increase in electron transport or a decrease in the electron path distance strategy was reported [41, 42]. Accordingly, increased charge collection efficiency in DSSC was reported by various research groups via incorporating graphene directly into the TiO_2 layer [37, 39, 43–52]. Tang et al. demonstrated a molecular grafting technique to fabricate the photoanode in which chemically exfoliated graphene sheets were integrated into the TiO_2 matrix. The graphene-implanted TiO_2 showed a PCE of 1.68%. Figure 1.5 gives a glimpse of the research work on graphene/TiO_2 photoanode application in DSSCs.

The incorporation of RGO helps create macropores inside the TiO_2 films, which serve as scattering centers for the incident sunlight and also can bind effectively with dye molecules to generate more photocurrent [53–54]. A systematic study on the influence of the graphene nanocomposite preparation by ex situ and the in situ route was reported by Nouri et al., where it was revealed that in situ doping, i.e., the addition of graphene oxide in the solution along with titania precursor was the most effective process for graphene integration [55]. The PCE can also be enhanced by employing co-sensitized nanocomposite thin films as photoanodes [56]. The obtained PCE as high as 9.45% was ascribed to high light absorption and increased LHE.

Recently, graphene quantum dots (GQDs) have also been demonstrated to be efficient in advancing the efficiency of DSSCs. Subramania and co-workers fabricated a nanocomposite photoanode using GQDs decorated on the surface of TiO_2 nanofibers [57]. The improved J_{sc} and V_{oc} resulted in PCE of 6.22% compared to 4.8% of the control cell. The increased photogenerated electron transfer ability, reduced charge recombination, increased dye intake, and significant LHE were obtained by the addition of QGD in TiO_2. Photoanode made of nitrogen-reduced GO increased the open-circuit voltage of the DSSC device. The maximum short circuit current and PCE are obtained when 0.2 wt% N-RGO has

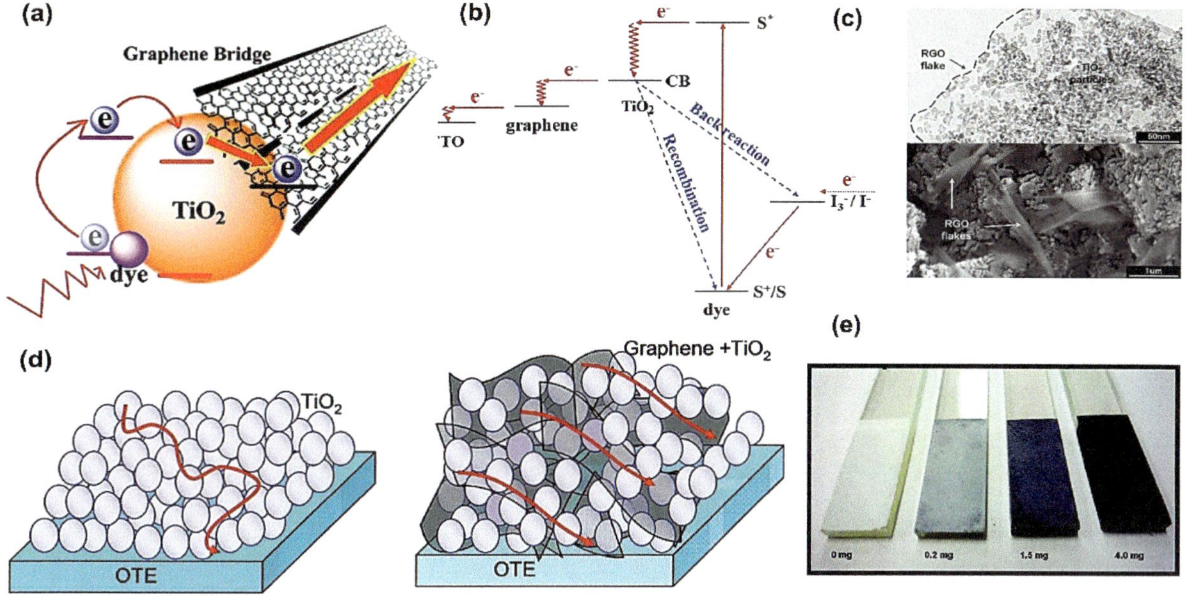

Figure 1.5 (a) Schematic illustration of 2D graphene bridges improved photoinduced charge transport in DSSCs [43]. (b) The electron transport process by 2D graphene bridges as an electron acceptor and transfers the electrons quickly with less recombination and backreaction [43]. (c) TEM & SEM image showing large RGO flakes scattered among the RGO–TiO_2 composite [49]. (d) Electron transport in pristine TiO_2 and RGO–TiO_2 nanocomposite on an optically transparent electrode (OTE) [44]. (e) Photo-graphs of TiO_2 thin films with different graphene loadings [44].

been integrated into TiO_2 [58]. Apart from TiO_2, ZnO and SnO_2 semiconductor oxides with graphene-incorporated photoanode were successfully developed for efficient DSSCs [59–61]. Thus, graphene materials were proved to play various roles when these are incorporated with semiconductor oxides and applied in DSSCs.

1.4.2.4 As a counter electrode

The counter electrode (CE), part of the DSSC, catalyzes the redox species completing a self-sustaining cell by transferring to the redox mediator electrons from the external circuit. A good and efficient CE having good conductivity and catalytic activity is necessary for a high-performing DSSC. Pt is most effectively used; however, it is undesirable to use the expensive Pt. Moreover, it also requires a high temperature for activation. Hence, materials that can replace Pt have been under exhaustive and extensive research including carbon materials, especially graphene, due to their fascinating properties such as good electrical and thermal conductivity, promising mechanical strength, and good optical transparency [62].

The first report on graphene as a CE for DSSC was by Hong et al. in 2008 [63]. A 60 nm thick composite CE film of graphene and polystyreneslufonate-doped poly(3,4-ethylenedioxythiophene) (graphene/PEDOT-PSS) conducting polymers were deposited onto ITO by spin coating at ambient temperatures. The presence of 1 wt% graphene displayed high transmittance (>80%) at visible wavelengths with high electrocatalytic activity and a PCE of 4.5%. Afterward, many reports on graphene and its composites came, inspired by the material's high surface area, high conductivity, and tunability. A partially-reduced graphene oxide (GO) film exhibited a PCE of 1.6% [64]. Hasin et al. studied the dependence of electrocatalytic properties of graphene films with different preparation conditions [65]. Later, Roy-Mayhew et al. observed that a thermally reduced GO porous network could be a sustainable competitor for platinum by effectively tuning the functional groups and defects, and found that the C/O ratio had a strong impact on the catalytic activity [66]. The functional graphene sheets (FGSs) as CE materials demonstrated a promising PCE of 5.5%. Further, a variety of graphene-based CE materials have been intensively developed for Pt-free DSSCs.

Graphene as a potential alternative to Pt as a counter electrode material in the DSSC was reported by the Gratzel group for the purpose of regenerating I^{3-}/I^- via an ionic solvent. However, DSSC using Pt-based counter electrodes still exceeds the efficiency (5.7%) with the graphene-based recording just 4.4% [67]. However, the same group also performed a study where the redox mediator was replaced with Co^{2+}/Co^{3+} using graphene nanoplatelets (GNP). It was observed that the GNP in the form of a thin semitransparent film graphene-based cathode system surpassed the traditional Pt CE-based system with 9.3% PCE [68].

Graphene has been used in various modifications to further enhance the PCE in the DSSC. Nagavolu et al. synthesized Pt-free solar-reduced graphene oxide having low charge transfer resistance for high PCE [69]. Xue et al. synthesized 3D nitrogen-doped graphene foams and observed a 7.07% PCE as a viable approach for preparing the metal-free catalyst [70]. Other groups adopted modifications such as loading with the metal for enhanced functionality. For example, Lu et al. used an iron-loaded graphene-based catalyst (Fe@3Dgraphene) for synthesizing ammonia [16]. Wei et al. synthesized cauliflower-fungus-like 3D graphene to understand the effect of structure and morphology. They utilized it for the replacement of the Pt counter electrode with 8.1% PCE compared to 7.5%. This value was about 10 times higher for a DSSC when only graphene was used [71]. Other graphene forms, such as nanogel, nanoplatelets, and nanoflakes, have been explored for use as CE in DSSC [62, 72, 73]. For example, the Li group synthesized graphene organogel of high porosity and interconnected network having high-specific area and conductivity with a 7.2% PCE [74].

Most often, researchers have used graphene or modified forms of graphene by using graphene as skeletons to dope and add semiconductors materials to gain excellent chemical and mechanical stability with surface area and porosity and good electrical conductivity. For example, Yao et al. wrapped mesoporous flower-like Fe_3O_4 onto nitrogen-doped graphene and achieved 9.26% PCE. They deduced a synergistic interaction of both materials to give an enhanced photoelectric property [75]. Selenium-doped graphene (SeG) prepared via ball milling, and high-temperature annealing has active sites responding actively to the reduction of I_3^- and showed PCE conversion of 8.42%.

The researchers used DFT calculations to further confirm that Se doping resulted in increasing charge transfer at the interface and reducing the ionization energy as well [76]. Other composites that can be explored for graphene include MN_4/graphene nanosheet to increase the catalytic activity, where M = Mn, Fe, Co, Ni, or Cu. For example, Cui et al. designed Co-based CoN_4/graphene with high activity. When used as CE, it was found to have high stability for the redox activity of I^-/I_3^- [77]. A DFT-based study proved that Co presents on the basal planes possesses excellent stability and catalytic activity for the reduction of triiodide. Also, the Co-N showed a slight mismatch compared to the other elements and binding of Co metal, and I further contributed to the activity of the material. Additionally, the Hao group synthesized amino-functionalized graphene by developing active sites on the surface of graphene [78].

Apart from the doping with semiconductor materials mentioned above, another way to increase the catalytic activity in graphene is to generate structural and edge defect sites. Additionally, graphene inherently suffers disadvantages because of its compatibility with polar electrolytes. This is addressed by introducing nitrogen on the carbon framework whereby the lone pairs in nitrogen (pyridinic and pyrrolic) activate sites to generate redox mediators. Furthermore, the surface polarity due to the electronegativity difference of N and C provides better surface wettability when used in polar electrolytes. For example, Fig. 1.6 (a and b) shows the SEM images of loosely packed microwave exfoliated graphene (MEG) and nitrogen-doped MEG (N-MEG) with the latter showing a densely packed structure relatively. Thus, the observed N-MEG structure showed a network of interconnected sheets, which are the curved and enhanced presence of active sites along the edges and also the surface porosity for easy transport of charge and ions [80]. This, in turn, contributes to increasing the catalytic activity of the graphene-based material. Additionally, Fig. 1.6 (c and d) shows TEM images with the MEG observing a flat surface compared to the wrinkling and crumpling surface of the N-MEG. The inverse fast Fourier transform (FFT) of N-MEG in Fig. 1.6 (e and f) confirms the hexagonal pattern in the material. Furthermore, the inverse FFT sees a larger part of N-MEG devoid of structural defects that can be further reached via a solution-based system of N-doping.

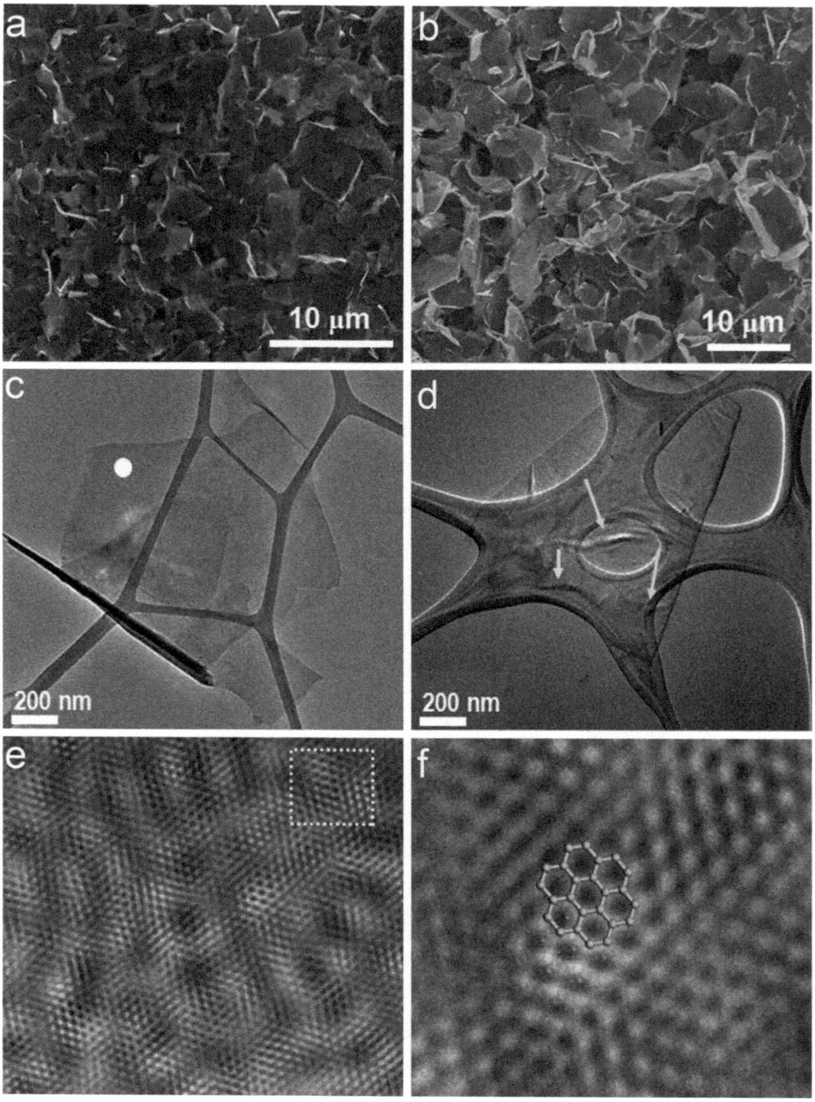

Figure 1.6 SEM images of (a) MEG and (b) N-MEG films on FTO. TEM images of a (c) flaky MEG nanosheet and (d) curved N-MEG nanosheet, the white arrow shows the apparent wrinkle. (e) The inverse FFT of the square region, and (f) a high-magnification image of the inverse FFT [79].

Hybridizing nanostructured transition metal sulfides/selenides with graphene are found to be very promising. Graphene acts as a matrix to prevent the self-aggregation of inorganic materials, offering copious pores to promote mass transfer and unified electronic conductive pathways to enable electron transfer. Many composites of transition metal sulfides and selenide composites with reduced GO have shown enhanced PCE. Ultrathin-walled Co_9S_8 nanotube/RGO nanocomposites, sandwich-like hierarchical structure CoS_2/RGO, and 2D $CoS_{1.097}$ sheets decorated with RGO nanosheets, etc. showed promising efficiency comparable to conventional Pt CE [81–83]. Another efficient CE reported was based on MoS_2/$CoSe_2$/$NiSe_2$ graphene nanocomposite [84–87]. In comparison with NiS and CoS, $NiCo_2S_4$ thiospinel has shown greater conductivity, activity, redox reactions, valence interchange, and charge hopping, compared to those of single cobalt or nickel sulfide. This improved performance was due to the coexistence of cobalt and nickel with different valences. DSSC fabricated with interconnected $NiCo_2S_4$ nanosheet CE exhibited a higher PCE of 7.22% in comparison with Pt-sputtered CE (6.87%), owing to exceptional charge transport characteristics and improved electrocatalytic activity [88]. Another interesting report on the growth of nanocrystalline $NiCo_2S_4$ by the microwave-assisted solvothermal method has shown good PCE in DSSCs (7.89%) in Fig. 1.7 [89]. The nanocrystallite aggregates of $NiCo_2S_4$/GNS nanohybrid unveiled a significant electrocatalytic activity (Fig. 1.7) toward triiodide (I^-/I_3^-) reduction, owing to the upgraded electronic conductivity and synergistic effects at $NiCo_2S_4$ and GNS interfaces.

Thus, it can be concluded that graphene and modifications of graphene, including doping with metals and non-metals, can improve the catalytic performance of graphene by altering the surface properties and chemical interactions in a 3D network. This further increases the electrical conductivity, and charge and ion transport in the system. Some of these synthesis techniques might be viable for mass production for use in commercial applications.

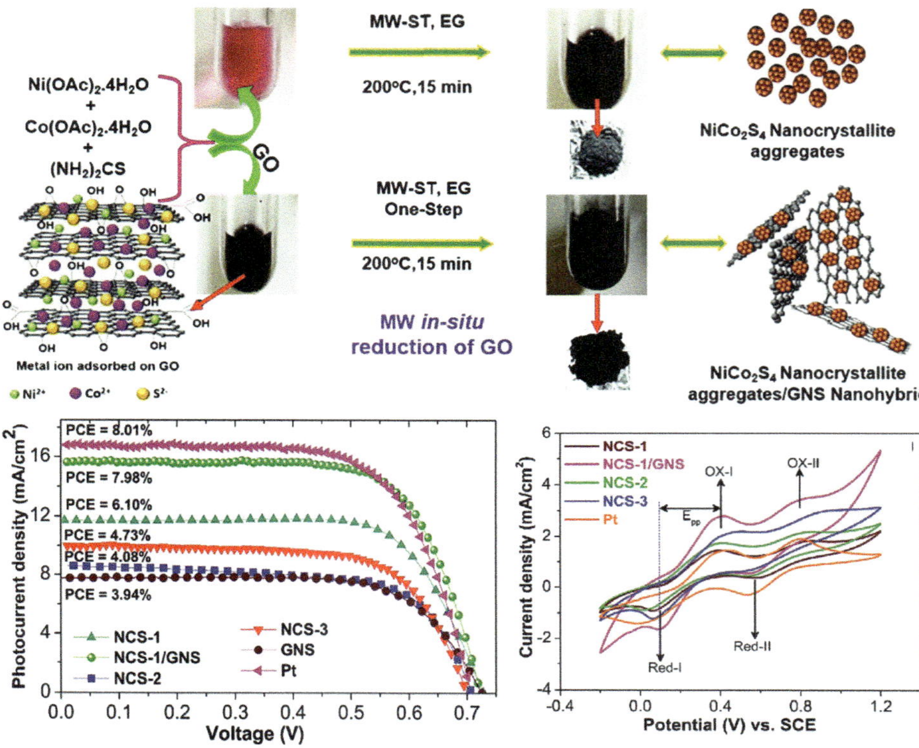

Figure 1.7 Schematic illustration of the reaction of a one-step microwave process for the synthesis of $NiCo_2S_4$ nanocrystallite aggregates prepared at a low temperature of 200 °C within 15 min. A comparative plot of J–V curves of DSSCs fabricated with different morphologies of $NiCo_2S_4$, the $NiCo_2S_4$/GNS hybrid, Pt, and pristine GNS counter electrodes. Cyclic voltammetry analysis of iodide/triiodide redox species for different morphologies of $NiCo_2S_4$, the $NiCo_2S_4$/GNS hybrid, and Pt CEs with a measured scan rate of 50 mV S^{-1} [89].

1.4.2.5 As an electrolyte

The primary role of an electrolyte present in a DSSC device is to act as a charge transport carrier between the photoanode and the counter electrode. In particular, the electrolyte plays a crucial role in dye regeneration, as a redox mediator, and as a conducting medium with the electrode interfaces, which can be either a solid or liquid medium. Initially, liquid electrolytes (LE) with different redox couples are used to fabricate DSSCs to facilitate charge carrier transport and regeneration of photosensitizing dye molecules. However, LE is readily volatile and associated with durability and stability issues. These can be solved by substituting a polymer or gel electrolyte with the LE. Therefore, solid graphene-based materials are desired to integrate into LE during the fabrication of DSSCs to attain high stability and better performance. A schematic representation of the electron transfer process in a graphene-incorporated LE-based DSSC is shown in Fig. 1.8.

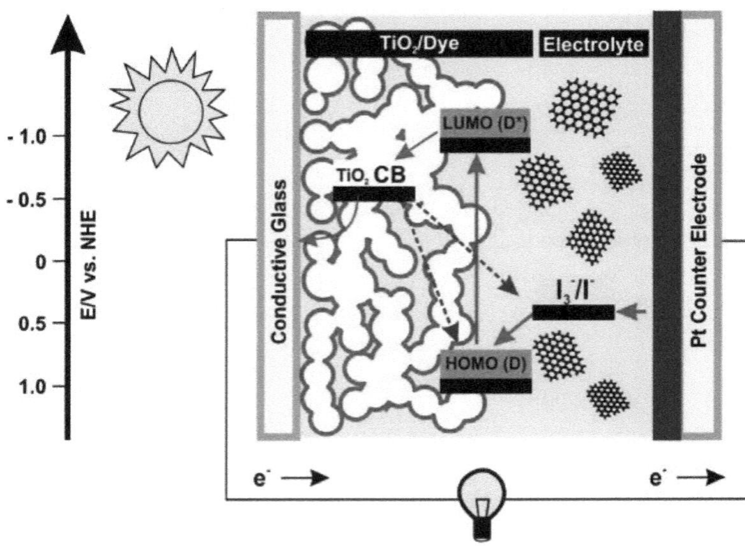

Figure 1.8 Schematic illustration of DSSC with graphene-incorporated liquid electrolyte [90].

Velten et al. reported the incorporation of few-layer graphene nanoribbons as an additive to the iodine-based electrolyte in an optically inverted DSSC architecture. They observed the improvement in photocurrent by ~2.2 mA/cm^2, which revealed that the effect of electrolyte bleaching plays a significant role in improving the PCE [91]. Ahmed et al. constructed DSSCs with graphene and a mixture of graphene and carbon nanotube (CNT) incorporated with 1-methyl-3-propylimidazolium iodide (PMII) ionic liquid-based quasi-solid-state electrolytes [92]. This study revealed that graphene-based quasi-solid-state electrolytes exhibit a higher percentage of light conversion efficiencies (2.10%) than pure PMII electrolytes (0.16%). The significant improvement in the overall efficiency is due to the graphene-based electrolyte, which served simultaneously both as charge transport material in the ionic liquids as well as a catalyst for the triiodide electrochemical reduction. Likewise, an iodide-functionalized graphene electrolyte was also applied for highly efficient DSSCs [93]. The study aimed to allow perfect contact between the surfaces of the sensitized porous TiO_2 photoanode and the graphene to function as the carrier mediator, which facilitates an effective electron transfer.

Brennan and his co-workers studied the effect of graphene nanoflakes incorporation into the ionic liquid electrolytes for DSSCs and achieved excellent PCEs as well as stability [94]. Up to 3 wt% of graphene nanoflakes were suspended in a 1-propyl-3-methyl imidazolium iodide (PMII) ionic liquid. As high as 20% increment in PCE compared to the plain ionic LE was observed. The authors claimed the significant improvement in PCE due to the graphene present in the viscous electrolyte, which acts as an efficient charge transfer material as well as a catalyst for the triiodide reduction in the electrolyte layer. Additionally, the self-organized networks formed between IL and graphene interactions promote efficient electron transfer in quasi-solid-state electrolytes. Later, the graphene-tailored polymer gel electrolytes were reported for quasi-solid-state DSSCs with 9.1% efficiency [95]. The presence of conductive graphene channels within the electrolyte increased the catalytic capability with reduced charge-transfer length resulting in superior charge kinetics.

Moreover, eco-friendly nonvolatile IL-based electrolytes were prepared from IL-GO and PMII for use in DSSCs by Lin and his co-workers [96]. Among the various percentages of IL-GO concentrations, 4 wt% of IL-GO addition achieved a maximum PCE of 7.04%. The increased value of PCE of 4 wt% is possibly due to the higher diffusion coefficients of triiodide, which increased the natural dye regeneration. The higher addition of IL-GO concentration (above 4 wt%) increases the viscosity of the electrolyte medium. This leads to difficulty in infiltration on the TiO_2 layer and also blocks the charge transfer kinetic reaction. This study explained that the appropriate quantity of IL-GO composite significantly improved the conductivity, the open-circuit voltage, the short-circuit current density, and the overall PCE of the DSSC device. The intensity-modulated photocurrent spectroscopy also revealed that 4 wt% of IL-GO electrolyte-based DSSC devices effectively collect more electrons before they recombine, which increases the redox reaction and improves photocurrent.

In another remarkable study, Chan et al. prepared a hybrid polymer gel electrolyte containing polyacrylonitrile (PAN) with graphene for quasi-solid DSSCs [97]. The inclusion of homogenous exfoliated 2D graphene into the PAN matrix dramatically reduced the ohmic contact to increase the electron transfer and also increased the diffusion coefficient of triiodide in the redox electrolyte. Yuan et al. also prepared to conduct gel electrolytes containing graphene-incorporated polyacrylate–poly(ethylene glycol) (PAA–PEG) and applied as an efficient electrolyte in DSSCs with remarkable efficiency of 7.74% [98]. This enhancement in PCE benefited from the interconnected conducting channels resulted from the rearrangement of graphene. The 3D framework structured gel electrolyte increased the electrocatalytic activity while decreasing the overpotential, thereby improving the open-circuit voltage. The same group also studied graphene incorporated into microporous poly(acrylic acid)–cetyltrimethylammonium bromide (PAA–CTAB) conducting gel electrolytes for DSSCs [99]. The resultant conducting gel electrolytes incorporated with graphene, graphene oxide, and nano graphite exhibited PCE of 7.06%, 6.35%, and 6.17%, respectively. The improved conductivity of graphene facilitated to conduct of reflux electrons from CE into its 3D matrix and was significantly involved in the I^-/I_3^- electrocatalytic redox reaction. Similarly, graphene-incorporated polymer electrolytes of PAA–PEG gel electrolyte, polyethylene glycol, polymethyl methacrylate, and poly(ethylene oxide) (PEO) have been studied for DSSC application and achieved enhanced PCE performances with stability [90, 95, 100–102].

Li and his co-workers prepared stacked graphene platelet nanofibers dispersed in the liquid electrolyte that proficiently increased the charge transport and catalytic reactions of the Co(III)/(II) redox couple observing PCE of 9.81% [103]. These can effectively increase the electrical conductivity of the device as well as provide more passage channels to accelerate fast charge transport. Recently, Venkatesan et al. reported highly efficient printable polymer electrolytes for DSSCs using GO sponge (GOS) with enhanced ion diffusivity and conductivity [104]. An excellent PCE of 8.78% and long-term stability were obtained with 1.5 wt% GOS. The hierarchical 3D structure of GOS with amphiphilic surfaces exhibited high surface area, large pore volume, and excellent stability. The aforementioned intrinsic properties of GOS can effectively improve both the efficiency and stability of the fabricated DSSC device [105].

1.4.2.6 As a photosensitizer

In DSSC, the photosensitizer is one of the main crucial components, which plays a vital role in harvesting sunlight. In the past two decades, various studies for photosensitized dyes have been investigated. Among the most studied photosensitized dyes, Ru-based photosensitizer is the most commonly studied because it absorbs sunlight in a wider spectral range (400–920 nm) of the solar spectrum. Additionally, it exhibited a longer lifetime and a higher number of electrochemical redox cycles, which are beneficial for a longer working DSSC device. DSSCs with graphene-based materials as photosensitizers exhibited reasonable improvements in PCE [106]. The desirable light-harvesting properties, tunable bandgap, sizeable optical absorption, and optical transparency are essential parameters for an effective light-harvesting sensitizer, and these properties can be well-adopted by carbon chemistry. The significant charge mobility in a graphene-based material can increase the charge collection in solar cells, thereby increasing the efficiency. Yan et al. prepared a uniform size of black graphene quantum dots (GQDs) and used it as a photosensitizer in DSSCs [106]. The less conjugated carbon atoms consisting of GQDs are readily soluble in appropriate organic solvents such as tetrahydrofuran, chloroform, and toluene. The properties, such as the broader light absorption spectrum and significant molar extinction coefficients of GQDs, are attractive as a potential material for solar cell applications. For example, GQDs in the dichloromethane absorption spectrum exhibit an absorbance maximum at 591 nm with the HOMO (5.3 eV) and LUMO (3.8 eV) parameters, which provide high support to use as a sensitizer in DSSCs. Moreover, the overall experiment revealed that GQDs as a sensitizer in DSSCs exhibit higher PCEs.

Kentrick et al. prepared GQD chemically (carboxyl linkers) anchored onto a single crystal TiO_2 (110) surface and used it as a sensitizer in the DSSC device. The experimental results revealed that the ultrafast electron injection from photoexcited GQDs to the TiO_2 CB with a time constant of $\tau_i < 15$ fs and had electron-hole recombination by fast and slow $\tau_{r1} = 80\text{–}130$ fs, $\tau_{r2} = 0.5\text{–}2$ ps, respectively [107]. These GQDs are further used as sensitizers in DSSCs and observe an enhancement in the overall PCE of DSSC devices. The hybrid DSSCs of TiO_2/GQD/N_3 dye fabricated by Mihalache et al. achieved excellent PCE. In hybrid DSSCs, the recombination redox electron is strongly hindered at the interface of the TiO_2/electrolyte interface by the GQDs [108]. In this study, GQDs co-sensitizer plays a crucial role during photoexcitation. This boosted the charge separation and collection, owing to the cascaded configuration of the energy levels. The photoluminescence property of GQDs also aids in the transfer of energy from GQDs to N3 dye, reducing the electron recombination by inhibition of back electron transfer. Esteban and co-workers prepared anthocyanin dye co-adsorbed with graphene as a sensitizer for DSSCs [109]. This study revealed that the anthocyanin dye co-adsorbed with graphene sensitizer showed decreased charge transfer resistance and observed a 2.4-fold increase in PCE of the DSSC devices. The excellent improvement can be attributed to the use of graphene, providing additional conducting pathways for the photogenerated electrons and significantly increasing PCE. Moreover, graphene present in the dye provides a longer electron lifetime in comparison to those without graphene photovoltaic cells. This longer electron lifetime indicates that graphene present in the DSSC cell can quickly suppress electron recombination in the electrolyte/dye interface.

Mehmood et al. also prepared graphene co-sensitization onto TiO_2 nanocomposite thin films with ruthenium sensitizers and metal-free organic photosensitizers for improving the PCE of DSSCs [56]. TiO_2/0.1%GR-N719/RK-1 sensitizer-based DSSCs exhibited a higher PCE of 9.45% than the sample without graphene. The higher PCE of the DSSCs can be attributed to improved light absorption and LHE. The addition of graphene offered an effective electron transfer pathway to the external circuit, which, in turn, improved the current density and charge recombination resistance. Gatti et al. used D-π-A (triphenylamino (D) – thiophene (π) – cyanoacrylic acid (A)) organic dye – RGO covalent dyad as a new photosensitizer for DSSC applications [110]. The D-π-A unit, which is covalently attached to the RGO surface, was investigated as a DSSC photosensitizer and achieved excellent PCE.

The continuing research and studies in this area will pave the way for new functional materials to be used as photosensitizers and act as efficient visible absorbers to promote the transfer of charges when irradiated with sunlight. The use of graphene in this frame will generate hybrid materials for the conversion of photons to electrons with high PCE with enhanced stability with the ambient environment. Additionally, graphene has proven to be excellent anchoring support with TiO_2 surfaces, as it exhibits less desorption tendency relative to other molecular sensitizers. Further understanding and development of new architectures and configurations such as bilayer hybrid can lead to improvement in the sensitizer applications.

1.4.3 Graphene in PSCs

In PSCs, graphene is used either as an n-type ETL or a p-type HTL. Generally, the HTL is deposited directly on an ITO or FTO substrate using liquid or vapor processing techniques. However, these expensive substrates are made of less earth-abundant materials and suffer from critical limitations such as poor mechanical properties and toxicity [111]. Graphene was demonstrated as a highly efficient nanoelectrode, owing to its high conductivity (low sheet resistance), flexibility, and high transparency of 97.4% with 100 Ω/sq sheet resistance.

1.4.3.1 As an ETL

An ETL plays a central role in PSCs. In order to fabricate a good-quality PSC device with excellent stability, charge transport layers should have good surface properties and should be compatible with the perovskite layer [112]. Also, it should provide better electron mobility, block holes, and avoid the back transfer of electrons. Furthermore, these materials should have the lowest

LUMO and the highest HOMO layers so that they must be perfectly band aligned with the absorbing perovskite layer. Metal oxides have been used as a general ELT as they satisfy these criteria. TiO_2 has been widely employed as an ETL due to its wide bandgap, high chemical stability, and suitable energy levels compatible with light-absorbing layers such as dyes, quantum dots, and perovskites [113–114]. Low-temperature processibility and good electrical properties also enabled these metal oxides to perform as good ETLs. Recombination losses can be effectively avoided, and electron mobility was improved by using carbonaceous materials including carbon nanotubes, graphene, and recently explored graphdiyne. Lately, graphene-based ETLs have been substantiated to be a good candidate for easy fabrication processes and to accomplish higher carrier injection with respect to the most frequently used ETLs [115]. As graphene shows remarkably high mobility and electronic and thermal conductivity, it has been incorporated into an ETL for PSC applications. Upon introducing graphene in the TiO_2 compact layer, the mobility of the ETL has been enhanced more than eight times compared with the pristine TiO_2.

Nicholas and co-workers introduced graphene for the first time as an electron-collecting layer in meso-superstructured PSCs to avoid high-temperature fabrication processing [116]. The nanocomposite of graphene and TiO_2 nanoparticles (NPs) were demonstrated as ETL instead of the commonly used mesoscopic TiO_2 ($mTiO_2$). The graphene nanoflakes in the composite offer superior charge-collection, facilitating the device fabrication at temperatures as low as 150 °C. The fabricated PSCs showed remarkable PV performance with a PCE of up to 15.6%. Enhancement in the PCE was ascribed due to the improved V_{oc}, J_{sc}, and FF, which is the result of synergetic effects between graphene and TiO_2 NPs. Later, Jung et al. investigated the influence of the amount of RGO on the photovoltaic performance of PSCs by electrochemical impedance spectroscopy and transient measurements [117]. The RGO/mp-TiO_2 nanocomposite as the ETL showed the best PCE of 14.5% by an effective reduction in interfacial resistance, thereby improvement in charge collection efficiency (i.e., enhanced electron transport and reduced charge recombination), as shown in Fig. 1.9 (a).

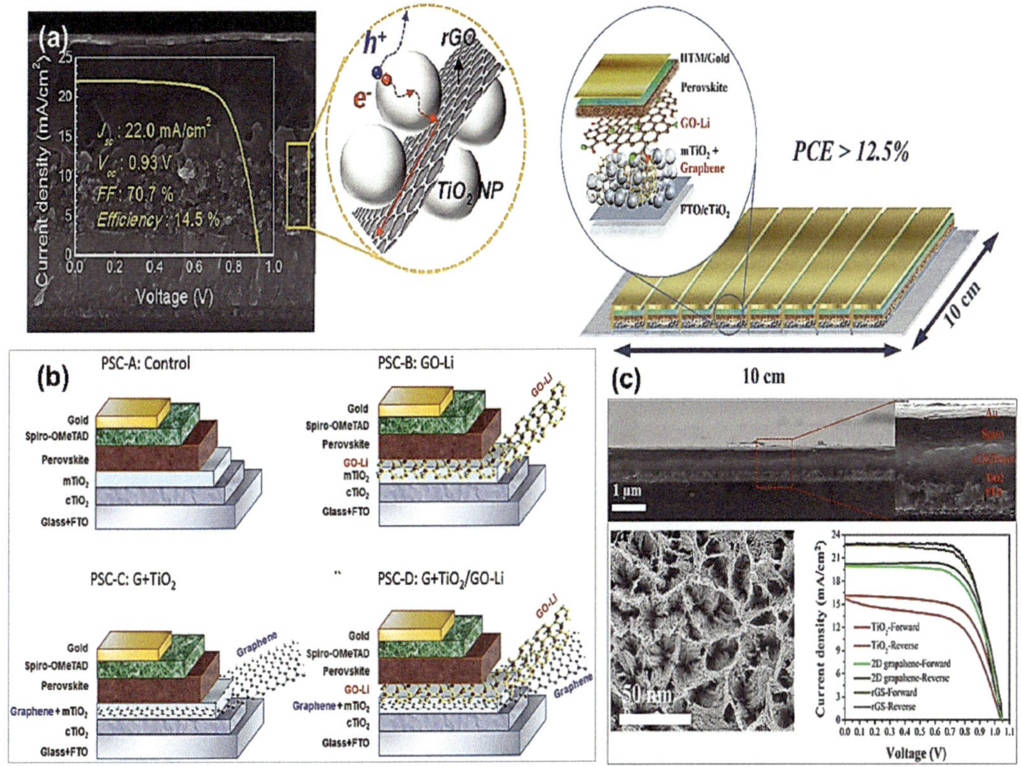

Figure 1.9 (a) Highly efficient PSCs with graphene-based ETLs, (b) PSC device architecture, and (c) cross-sectional SEM images and PV performance of devices [117–119].

Lithium-neutralized graphene oxide (GO-Li) has proved to be promising for effective electron extraction from the perovskite layer [120]. Replacement of H atoms of GO by Li atoms can effectively reduce the working function (WF) of GO from 4.9 to 4.3 ± 0.1 eV. Additionally, the low electronegativity and low WF of Li result in the transfer of valence electrons to the GO plane, the formation of Li^+ dipoles, and finally, a shift in the Fermi level toward the vacuum and a consequent decrease in WF. This is highly desirable as it is a good match with the LUMO level of $mTiO_2$. The fabricated devices with Li-modified graphene exhibited enhanced J_{sc} and FF with reduced work function and hysteresis comparable to the device with pristine TiO_2. Moreover, the GO-Li layer has demonstrated the ability to passivate the $mTiO_2$ oxygen defects/vacancies that are the principal reaction centers for moisture attack. Hence, a more stable and extended device lifetime under 1 SUN prolonged stress test was obtained. Another interesting work involved the insertion of graphene flakes into the $mTiO_2$ layer adds to the GO-Li for PSCs [118]. These interlayer structured PSCs with perovskite/$mTiO_2$ interface exhibited improved PCE and stability for both small (0.1 cm^2) and large areas (10 × 10 cm^2 modules with an active area of 6.32 cm^2). The schematic of device architecture is given in Fig. 1.9 (b). Biccari et al. studied in detail the carrier recombination dynamics in $CH_3NH_3PbI_3$ (MAPI) based PSCs [121]. When a graphene-doped mesoporous TiO_2 (G+$mTiO_2$) layer was used in addition to a GO-Li interlayer in ETL architecture, the morphology of the MAPI film embedded in the mesoporous layer was found to be frozen in the tetragonal phase, irrespective of the temperature treatments and the defect concentration is lowered one order of magnitude than other ETLs. An upsurge in the electron collection efficiency and the crystalline quality of the active material was obtained, which resulted in enhanced performance and stability of fabricated PSCs. The role of graphene flakes on the PCE and stability of PSCs was further confirmed by Carlo and co-workers using femtosecond transient absorption (TA) spectroscopy [119]. A simple chemical electrophoretic procedure was used to synthesize a 3D reduced-graphene scaffold (rGS) with superior surface area (Fig. 1.9 (c)) and utilized as an interface layer between perovskite and TiO_2

films. The improved carrier transportation by 27% with a PCE of 17.2% was obtained with high stability and low hysteresis. The presence of rGS lowered the work function, which facilitated the easy transfer of the photoinduced electrons to the rGS with better conductivity. Tavakoli et al. investigated the instability issue of the MAPI perovskite material for ZnO NPs-based ETL and reported that incorporated RGO acts as a protection layer to improve device stability [122]. The ZnO/graphene layer facilitated the extraction and transfer of electrons at the interface, which increased the EQE and the current density, owing to the quenching effect.

Studies reported that the graphene incorporation could improve the high trap-state density and current-voltage hysteresis of the SnO_2-based ETL layer for PSCs. Zhao et al. demonstrated the fabrication of highly efficient planar PSCs with PCE of 20.2% and FF of 82% by the incorporation of chemically modified naphthalene di-imide-graphene into the nanocrystal SnO_2-based ETL [123]. The highly conductive chemically modified 2D graphene layer increased the surface hydrophobicity. As a result of this, effective van der Waals interactions were formed between the surfactant and the organic-inorganic hybrid lead halide perovskite compounds, which increased the electron mobility and extraction ability, reduced carrier recombination, and improved the FF of PSCs. The device architecture and cross-sectional SEM images of the fabricated PSC are shown in Fig. 1.10 (a and b). Recent work reported integrating graphene ink prepared by liquid-phase exfoliation into SnO_2 ETL to yield a PCE of 18% with negligible hysteresis [124]. The fabricated PSCs showed high stability, with 90% of the initial PCE value after storing in ambient conditions with a relative humidity of 40 ± 5% for 300 h. The photovoltaic performance data of the fabricated device is shown in Fig. 1.10 (b).

Inverted planar PSCs also use graphene as ETL, and here the graphene is used in combination with PCBM and semiconductor oxide layer. The incorporation of RGO in PCBM ETL results in increased conductivity, which results in high J_{sc}, and FF while reducing the surface traps and passivates the perovskite surface [125]. As a result of this, a device with improved V_{oc} and reduced light soaking effect can be fabricated.

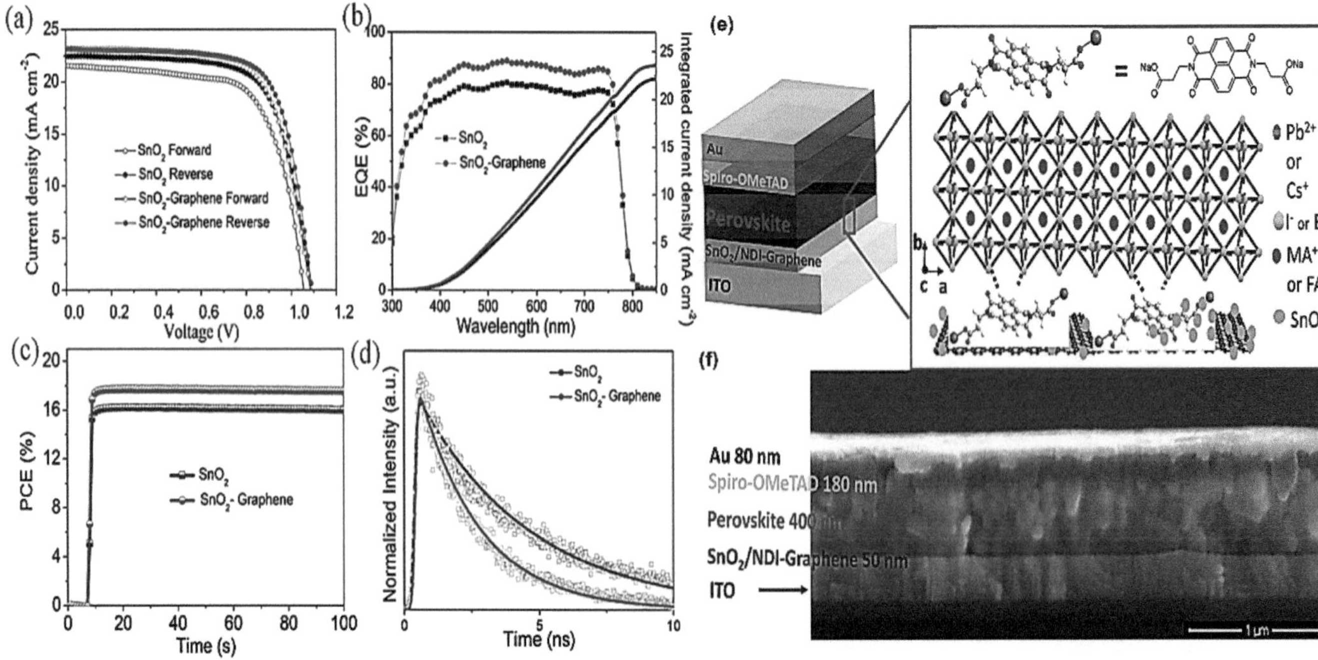

Figure 1.10 (a)–(d) Photovoltaic performance and luminescence spectra of SnO_2-based PSCs modified with graphene. (e) and (f) Device architecture and FESEM image of the PSC device [123–124].

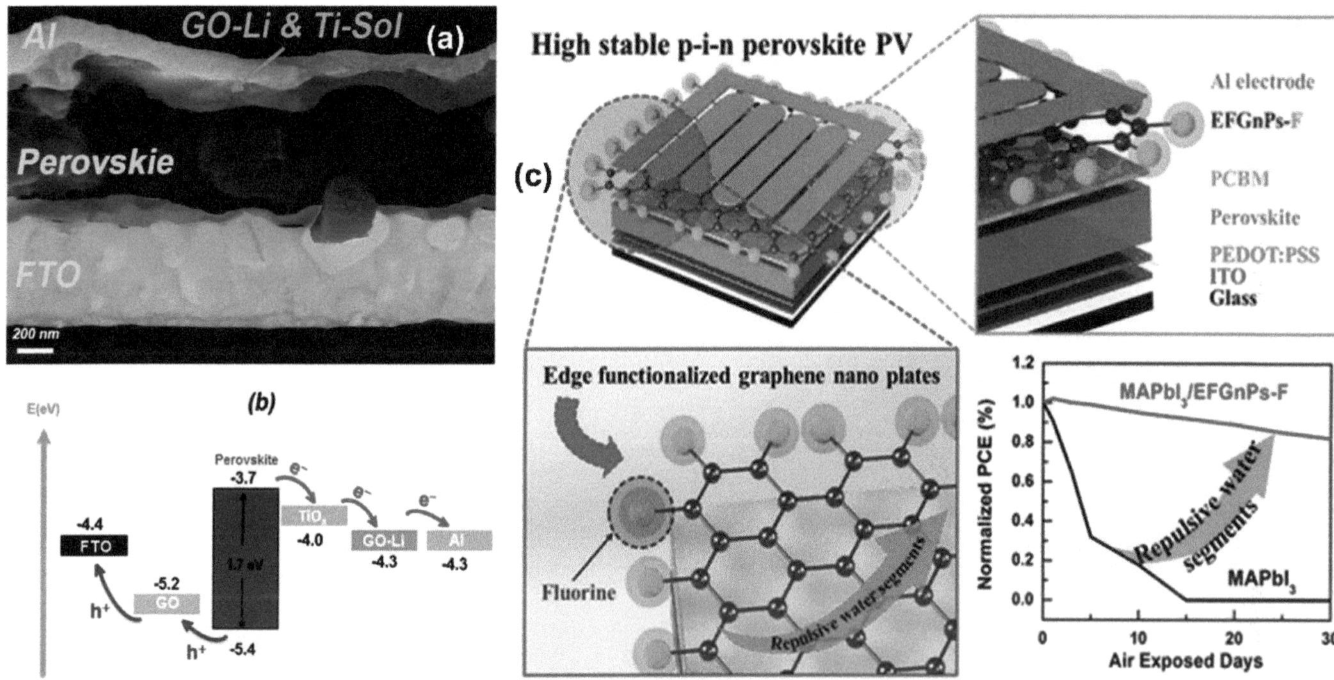

Figure 1.11 (a) Cross-sectional FESEM image of the subsequent layers of inverted planar PSCs with GO-Li and Ti-based sol as electron-transporting layers. (b) Energy level diagram showing possible charge transfer in the inverted planar PSC device. (c) Schematic illustration of the device architecture and stability measurement of inverted planar PSCs with fluorine functionalized graphene as ETL [128].

Furthermore, RGO was found to stabilize the PCBM/perovskite interface and, in turn, the perovskite crystal. Accordingly, Li-zunctionalized GO and fluorine (F) functionalized graphene nanoplatelets (EFGnPs-F) also showed improved performance in inverted planar PSCs [126–128]. A schematic illustration of charge transfer and the FESEM image showing the device architecture is given in Fig. 1.11. EFGnPs-F in PSCs demonstrated the prevention of air degradation, owing to the chemical inertness and hydrophobic properties offered by the C–F bond.

Thus, the graphene-based ETLs have proven to possess remarkable influence on the PSCs PCEs and the fabrication of the entire devices in comparison to TiO_2-based ETLs, high-temperature processing that makes PSCs versatile in their applications. These offer higher charge collection due to good conductivity and lesser series resistance at the interfaces and do not induce any parasitic absorption losses (as CdS does) that might lower the PSCs' performance. Graphene can operate as a super-fast bridge for electron transfer from the perovskite absorber to the ETL. Additionally, a variation of the RGO/mp-TiO_2 nanocomposite ETL with surface passivation can also increase the electrical properties of the composite ETL and thus enhance the PCE of PSC.

1.4.3.2 As an HTL

In a typical PSC, HTL plays an important role in achieving high PCEs by influencing the V_{oc} and stability of the device. The primary function of HTL is that it acts as a barrier between the anode and the perovskite layer to block the electron to the anode, and it facilitates the smooth passage of holes. The HTM layer can improve the surface coverage and act as a separating layer between the top contact and the bottom contact/transport layer, thus reducing the chances of charge recombination. The hole-transporting material used for efficient hole extraction should possess high hole mobility and low electron affinity to reduce recombination losses during transportation to the hole-collecting contacts. It should possess the necessary ionization potential to match the HOMO and LUMO levels of perovskite to minimize the injection losses. It should have high thermal stability and resistance to air and moisture. And finally, it should be earth-abundant and cost-effective.

As of now, various HTMs comprising inorganic (e.g., NiO, CuSCN, CuI, Cu_xO, and GO), organic (e.g., 2,2′,7,7′-tetrakis(N,N-di-p-methoxyphenylamine)–9,9′-spirobifluorene)-spiro-OMeTAD, DOR3T-TBDT, and Fused-F), polymers (P3HT, PEDOT: PSS, and poly(triarylamine) PTAA), and oligomers were applied in PSCs [129]. Among these, spiro-OMeTAD [2,20,7,70-tetrakis-(N,N-di-p-methoxyphenylamine) 9,90-spirobifluorene] is the most common HTM for normal mesoporous PSCs but it suffers from insufficient charge mobility. This problem can be resolved by doping in spiro material, but these dopants often reduce the stability of the HTL. In pursuit of resolving the stability issues, graphene can be employed as it can provide better hole extraction, film passivation, wettability for spiro-OMeTAD, and enhanced stability for the device.

Luo et al. reported the application of p-type, highly conductive RGO reduced by ferrous iodide for dopant-free spiro-OMeTAD HTL for PSCs [130]. The solution-processable RGO showed improved PV performance with superior stability and PCE. Similarly, amino-functionalized graphene (NGs) incorporated between the active and dopant-free spiro-OMeTAD layer showed a 36% improvement in PCE related to the graphene-free sample [131]. Here, the fabricated device showed reduced trap states, owing to the improved hole capture as well as the transfer pathway provided by RGO. The complete replacement of spiro-based HTL from the PSC by well-defined thiolated nanographene is another interesting study [132]. The enhanced stability of the PSCs by applying RGO was further verified by Carlo and co-workers using the endurance tests over 1987 hours of shelf life and 120 hours of light soaking [133].

The application of a thin layer of graphene before the PEDOT:PSS deposition in inverted PSC architecture also provided superior device stability [134]. Doped graphene structures also showed improved performance as HTL. In a significant study by Kim et al., graphene doped with $AuCl_3$ was used as a protective layer between PEDOT:PSS and ITO [135]. The introduction of doped graphene aided to protect ITO and facilitate the collection of holes. Additionally, the stability of the fabricated PSCs was found to be considerably enhanced by maintaining 67% of its original PCE after 30 days. The details on the device architecture, the cross-sectional FESEM image, and the possible electron transfer mechanism can be understood from Fig. 1.12 (a)–(d). The facile spin-coated ammonia-modified graphene ($GO:NH_3$) layer over the deposited PEDOT:PSS layer also succeeded in obtaining the best PV performance in PSCs. A 2-nm thin layer of $GO:NH_3$ paves the way to the formation of an active perovskite film with improved crystallization and preferred orientation [136]. The alignment of different energy levels was found to be better matched at the perovskite interface, which is mainly responsible for the best PV performance (Fig. 1.12 (b)).

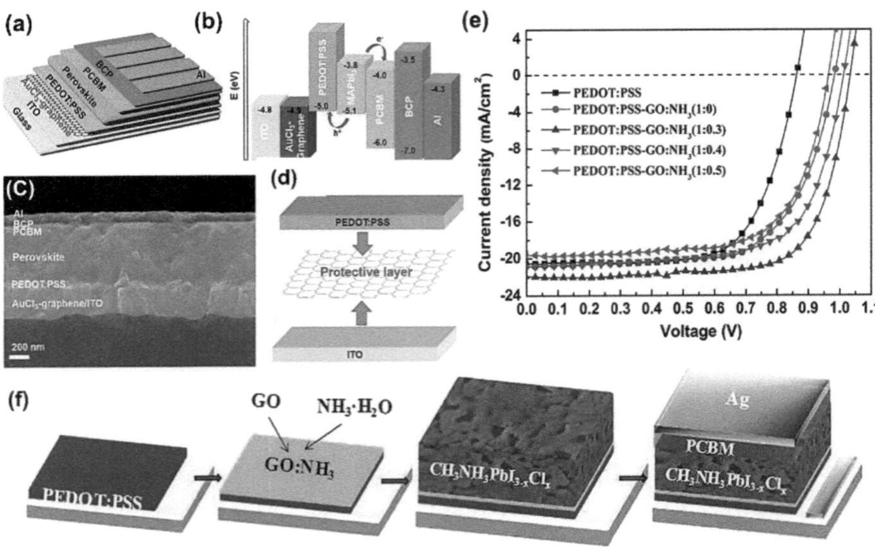

Figure 1.12 (a) and (b) Schematics of solar cell using $AuCl_3$-doped GR as a protecting layer and corresponding energy level diagram. (c) Cross-sectional SEM image of a PSC displaying the constituting layers. (d) A schematic diagram describing the GR as a protecting layer between PEDOT:PSS and ITO. (e) *J–V* curves of PSC devices with a PEDOT:PSS-$GO:NH_3$ HTL with different amounts of ammonia into GO aqueous solution (ratio of GO solution: ammonia = 1:0, 1:0.3, 1:0.4, 1:0.5) during the preparation. (f) Flowchart of fabrication procedure of PSCs with the PEDOT:PSS-$GO:NH_3$ layer [135].

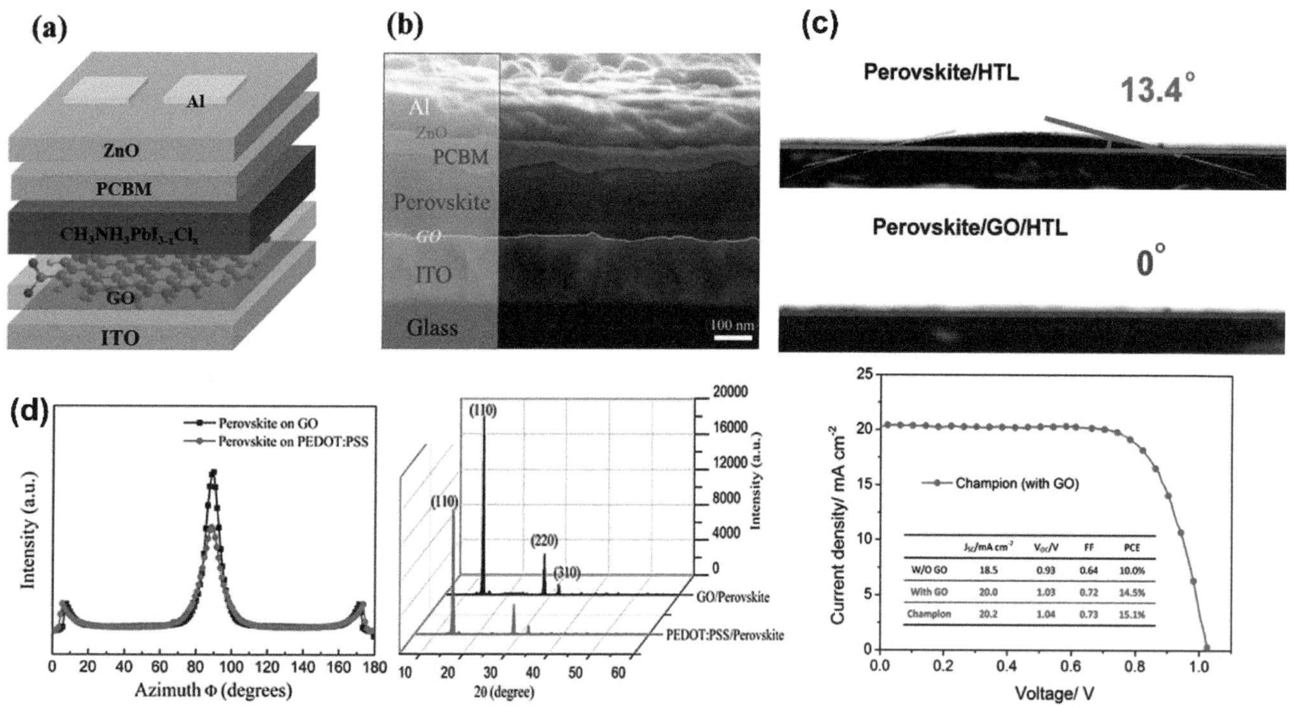

	J_{sc}/mA cm^{-2}	V_{oc}/V	FF	PCE
W/O GO	18.5	0.93	0.64	10.0%
With GO	20.0	1.03	0.72	14.5%
Champion	20.2	1.04	0.73	15.1%

Figure 1.13 (a) Representation of inverted solar cell configuration consisting of a structure of ITO/GO/ CH3NH3PbI$_{3-x}$Cl$_x$/PCBM/ZnO/Al. (b) Cross-sectional SEM image of the optimized inverted device configuration. (c) HTL solution contact angles in perovskite films with GO-free and with GO modification on FTO. (d) PV parameters of the GO-based PSC [137, 138].

Pristine GO also has been successfully used as a replacement for PEDOT:PSS HTL in PSCs. Wu et al. first introduced GO as a hole conductor (Fig. 1.13 (a)) and a PCE as high as 12.4% was obtained [137]. The underlaid GO layer, as seen in the FESEM image of Fig. 1.13 (b), efficiently extracted the hole out of the perovskite layer and enabled the formation of large homogenous domains, which is responsible for increasing the surface coverage. The study revealed that the $CH_3NH_3PbI_{3-x}Cl_x$ perovskite films grown on GO were highly textured with an in-plane orientation of the (110) plane (Fig. 1.13 (d)) of GIXRD. The properties like domain size, crystal orientation, and the morphology of the perovskite film with uniform coverage gave enhanced device performance. In a different study, GO was employed as a dual-function buffer layer at the interface among perovskite and spiro-OMeTAD (HTM), which decreased the charge recombination and contact angle from ~13.4 to 0 (Fig. 1.13 (c)) degree ensuing enhanced PV performance and stability [138].

The implementation of graphene as HTL in PSCs is significant since it improves the PV characteristics, due to the efficient charge extraction and transfer. The tuning of energy levels to achieve the desirable electron transfer as well as engineering of morphology and solubility, low cost, and environmentally friendliness make graphene the most suitable material for HTL applications. Additionally, graphene material provides superior stability to the fabricated devices, which helps in the commercialization of PSCs. The high-quality production of graphene is still challenging as CVD preparation is not preferred for the mass production of graphene.

1.5 Conclusion

This chapter systematically summarized the advancements in the application of graphene for highly efficient DSSCs and PSCs. With its distinctive advantages, graphene has proved the ability to perform various roles in solar cell architectures and led the way for the overall improvement of device performance. The current research direction focuses on the design and development of cost-effective and large-scale fabrication of flexible solar cells with highly reproducible performance using graphene as an efficient light-harvesting and exciton separation component. Even though vast development has been achieved, various concerns remain unexplored, like developing novel synthetic approaches to effectively blend single-layer graphene sheets with polymer matrices and transitional metals to get the maximum advantage of both materials. Also, there is a need to develop specific characterization techniques to find out the exact determination of reaction mechanisms taking place inside the device. This will increase the impact of graphene used in the device as the performance is currently evaluated through certain assumptions and the performance of devices. Additionally, more scientific efforts are needed for the fabrication and scaling up of graphene-based solar cells with high efficiency and long-term stability.

References

1. Gong, J.; Li, C.; Wasielewski, M. R., Advances in solar energy conversion. *Chemical Society Reviews* **2019**, *48*, 1862–1864.
2. Lewis, N. S., Introduction: Solar energy conversion. *Chemical Reviews* **2015**, *115*, 12631–12632.
3. Qazi, A.; Hussain, F.; Rahim, N. A.; Hardaker, G.; Alghazzawi, D.; Shaban, K.; Haruna, K., Towards sustainable energy: A systematic review of renewable energy sources, technologies, and public opinions. *IEEE Access* **2019**, *7*, 63837–63851.
4. Brown, G. F.; Wu, J., Third generation photovoltaics. *Laser & Photonics Reviews* **2009**, *3*, 394–405.
5. Nayak, P. K.; Mahesh, S.; Snaith, H. J.; Cahen, D., Photovoltaic solar cell technologies: Analysing the state of the art. *Nature Reviews Materials* **2019**, *4*, 269–285.
6. O'Regan, B.; Grätzel, M., A low-cost, high-efficiency solar cell based on dye-sensitized colloidal TiO_2 films. *Nature* **1991**, *353*, 737–740.
7. Nazeeruddin, M. K.; Baranoff, E.; Grätzel, M., Dye-sensitized solar cells: A brief overview. *Solar Energy* **2011**, *85*, 1172–1178.
8. Hagfeldt, A., Brief overview of dye-sensitized solar cells. *AMBIO* **2012**, *41*, 151–155.
9. J. Frank, A.; Kopidakis, N.; Lagemaat, J. v. d., Electrons in nanostructured TiO_2 solar cells: Transport, recombination and

photovoltaic properties. *Coordination Chemistry Reviews* **2004**, *248*, 1165–1179.

10. Omar, A.; Abdullah, H., Electron transport analysis in zinc oxide-based dye-sensitized solar cells: A review. *Renewable and Sustainable Energy Reviews* **2014**, *31*, 149–157.
11. Mathew, S.; Yella, A.; Gao, P.; Humphry-Baker, R.; Curchod, B. F. E.; Ashari-Astani, N.; Tavernelli, I.; Rothlisberger, U.; Nazeeruddin, M. K.; Grätzel, M., Dye-sensitized solar cells with 13% efficiency achieved through the molecular engineering of porphyrin sensitizers. *Nature Chemistry* **2014**, *6*, 242–247.
12. Selvaraj, P.; Baig, H.; Mallick, T. K.; Siviter, J.; Montecucco, A.; Li, W.; Paul, M.; Sweet, T.; Gao, M.; Knox, A. R.; Sundaram, S., Enhancing the efficiency of transparent dye-sensitized solar cells using concentrated light. *Solar Energy Materials and Solar Cells* **2018**, *175*, 29–34.
13. Zhou, D.; Zhou, T.; Tian, Y.; Zhu, X.; Tu, Y., Perovskite-based solar cells: materials, methods, and future perspectives. *Journal of Nanomaterials* **2018**, *2018*, 8148072.
14. Kojima, A.; Teshima, K.; Shirai, Y.; Miyasaka, T., Organometal halide perovskites as visible-light sensitizers for photovoltaic cells. *Journal of the American Chemical Society* **2009**, *131*, 6050–6051.
15. Jung, E. H.; Jeon, N. J.; Park, E. Y.; Moon, C. S.; Shin, T. J.; Yang, T.-Y.; Noh, J. H.; Seo, J., Efficient, stable and scalable perovskite solar cells using poly(3-hexylthiophene). *Nature* **2019**, *567*, 511–515.
16. Li, X.; Bi, D.; Yi, C.; Décoppet, J.-D.; Luo, J.; Zakeeruddin, S. M.; Hagfeldt, A.; Grätzel, M., A vacuum flash-assisted solution process for high-efficiency large-area perovskite solar cells. *Science* **2016**, *353*, 58.
17. Petrus, M. L.; Schlipf, J.; Li, C.; Gujar, T. P.; Giesbrecht, N.; Müller-Buschbaum, P.; Thelakkat, M.; Bein, T.; Hüttner, S.; Docampo, P., Capturing the sun: A review of the challenges and perspectives of perovskite solar cells. *Advanced Energy Materials* **2017**, *7*, 1700264.
18. Ansari, M. I. H.; Qurashi, A.; Nazeeruddin, M. K., Frontiers, opportunities, and challenges in perovskite solar cells: A critical review. *Journal of Photochemistry and Photobiology C: Photochemistry Reviews* **2018**, *35*, 1–24.
19. Mesquita, I.; Andrade, L.; Mendes, A., Perovskite solar cells: Materials, configurations and stability. *Renewable and Sustainable Energy Reviews* **2018**, *82*, 2471–2489.

20. Urieta-Mora, J.; García-Benito, I.; Molina-Ontoria, A.; Martín, N., Hole transporting materials for perovskite solar cells: A chemical approach. *Chemical Society Reviews* **2018**, *47*, 8541–8571.

21. Malinkiewicz, O.; Yella, A.; Lee, Y. H.; Espallargas, G. M.; Graetzel, M.; Nazeeruddin, M. K.; Bolink, H. J., Perovskite solar cells employing organic charge-transport layers. *Nature Photonics* **2014**, *8*, 128–132.

22. Novoselov, K. S.; Geim, A. K.; Morozov, S. V.; Jiang, D.; Zhang, Y.; Dubonos, S. V.; Grigorieva, I. V.; Firsov, A. A., Electric field effect in atomically thin carbon films. *Science* **2004**, *306*, 666.

23. Zhen, Z.; Zhu, H., 1-Structure and properties of graphene. In *Graphene*, Zhu, H.; Xu, Z.; Xie, D.; Fang, Y., Eds. Academic Press: 2018; pp. 1–12.

24. Allen, M. J.; Tung, V. C.; Kaner, R. B., Honeycomb carbon: A review of graphene. *Chemical Reviews* **2010**, *110*, 132–145.

25. Das, S.; Sudhagar, P.; Kang, Y. S.; Choi, W., Graphene synthesis and application for solar cells. *Journal of Materials Research* **2014**, *29*, 299–319.

26. Geim, A. K.; Novoselov, K. S., The rise of graphene. *Nature Materials* **2007**, *6*, 183–191.

27. Hummers, W. S.; Offeman, R. E., Preparation of graphitic oxide. *Journal of the American Chemical Society* **1958**, *80*, 1339–1339.

28. Marcano, D. C.; Kosynkin, D. V.; Berlin, J. M.; Sinitskii, A.; Sun, Z.; Slesarev, A.; Alemany, L. B.; Lu, W.; Tour, J. M., Improved synthesis of graphene oxide. *ACS Nano* **2010**, *4*, 4806–4814.

29. Zaaba, N. I.; Foo, K. L.; Hashim, U.; Tan, S. J.; Liu, W.-W.; Voon, C. H., Synthesis of graphene oxide using modified hummers method: Solvent influence. *Procedia Engineering* **2017**, *184*, 469–477.

30. Muzyka, R.; Kwoka, M.; Smędowski, Ł.; Díez, N.; Gryglewicz, G., Oxidation of graphite by different modified Hummers methods. *New Carbon Materials* **2017**, *32*, 15–20.

31. Wang, X.-Y.; Narita, A.; Müllen, K., Precision synthesis versus bulk-scale fabrication of graphenes. *Nature Reviews Chemistry* **2017**, *2*, 0100.

32. Major, S.; Chopra, K. L., Indium-doped zinc oxide films as transparent electrodes for solar cells. *Solar Energy Materials* **1988**, *17*, 319–327.

33. Patil, K.; Rashidi, S.; Wang, H.; Wei, W., Recent progress of graphene-based photoelectrode materials for dye-sensitized solar cells. *International Journal of Photoenergy* **2019**, *2019*, 1–16.

34. Wang, X.; Zhi, L.; Müllen, K., Transparent, conductive graphene electrodes for dye-sensitized solar cells. *Nano Letters* **2008**, *8*, 323–327.
35. Bi, H.; Sun, S.; Huang, F.; Xie, X.; Jiang, M., Direct growth of few-layer graphene films on SiO_2 substrates and their photovoltaic applications. *Journal of Materials Chemistry* **2012**, *22*, 411–416.
36. Kim, S. R.; Parvez, M. K.; Chhowalla, M., UV-reduction of graphene oxide and its application as an interfacial layer to reduce the back-transport reactions in dye-sensitized solar cells. *Chemical Physics Letters* **2009**, *483*, 124–127.
37. Wang, H.; Leonard, S. L.; Hu, Y. H., Promoting effect of graphene on dye-sensitized solar cells. *Industrial & Engineering Chemistry Research* **2012**, *51*, 10613–10620.
38. Chen, T.; Hu, W.; Song, J.; Guai, G. H.; Li, C. M., Interface functionalization of photoelectrodes with graphene for high performance dye-sensitized solar cells. *Advanced Functional Materials* **2012**, *22*, 5245–5250.
39. Durantini, J.; Boix, P. P.; Gervaldo, M.; Morales, G. M.; Otero, L.; Bisquert, J.; Barea, E. M., Photocurrent enhancement in dye-sensitized photovoltaic devices with titania–graphene composite electrodes. *Journal of Electroanalytical Chemistry* **2012**, *683*, 43–46.
40. Tang, B.; Hu, G., Two kinds of graphene-based composites for photoanode applying in dye-sensitized solar cell. *Journal of Power Sources* **2012**, *220*, 95–102.
41. Chappel, S.; Grinis, L.; Ofir, A.; Zaban, A., Extending the current collector into the nanoporous matrix of dye sensitized electrodes. *The Journal of Physical Chemistry B* **2005**, *109*, 1643–1647.
42. Joanni, E.; Savu, R.; de Sousa Góes, M.; Bueno, P. R.; de Freitas, J. N.; Nogueira, A. F.; Longo, E.; Varela, J. A., Dye-sensitized solar cell architecture based on indium–tin oxide nanowires coated with titanium dioxide. *Scripta Materialia* **2007**, *57*, 277–280.
43. Yang, N.; Zhai, J.; Wang, D.; Chen, Y.; Jiang, L., Two-dimensional graphene bridges enhanced photoinduced charge transport in dye-sensitized solar cells. *ACS Nano* **2010**, *4*, 887–894.
44. Ng, Y. H.; Lightcap, I. V.; Goodwin, K.; Matsumura, M.; Kamat, P. V., To What extent do graphene scaffolds improve the photovoltaic and photocatalytic response of TiO_2 nanostructured films? *The Journal of Physical Chemistry Letters* **2010**, *1*, 2222–2227.
45. Zhu, P.; Nair, A. S.; Shengjie, P.; Shengyuan, Y.; Ramakrishna, S., Facile fabrication of TiO_2–graphene composite with enhanced photovoltaic

and photocatalytic properties by electrospinning. *ACS Applied Materials & Interfaces* **2012**, *4*, 581–585.

46. Tang, Y.-B.; Lee, C.-S.; Xu, J.; Liu, Z.-T.; Chen, Z.-H.; He, Z.; Cao, Y.-L.; Yuan, G.; Song, H.; Chen, L.; Luo, L.; Cheng, H.-M.; Zhang, W.-J.; Bello, I.; Lee, S.-T., Incorporation of graphenes in nanostructured TiO_2 films via molecular grafting for dye-sensitized solar cell application. *ACS Nano* **2010**, *4*, 3482–3488.
47. Zhu, G.; Xu, T.; Lv, T.; Pan, L.; Zhao, Q.; Sun, Z., Graphene-incorporated nanocrystalline TiO_2 films for CdS quantum dot-sensitized solar cells. *Journal of Electroanalytical Chemistry* **2011**, *650*, 248–251.
48. Fang, X.; Li, M.; Guo, K.; Zhu, Y.; Hu, Z.; Liu, X.; Chen, B.; Zhao, X., Improved properties of dye-sensitized solar cells by incorporation of graphene into the photoelectrodes. *Electrochimica Acta* **2012**, *65*, 174–178.
49. Bell, N. J.; Ng, Y. H.; Du, A.; Coster, H.; Smith, S. C.; Amal, R., Understanding the enhancement in photoelectrochemical properties of photocatalytically prepared TiO_2-reduced graphene oxide composite. *The Journal of Physical Chemistry C* **2011**, *115*, 6004–6009.
50. Sun, S.; Gao, L.; Liu, Y., Enhanced dye-sensitized solar cell using graphene-TiO_2 photoanode prepared by heterogeneous coagulation. *Applied Physics Letters* **2010**, *96*, 083113.
51. He, Z.; Guai, G.; Liu, J.; Guo, C.; Chye Loo, J. S.; Li, C. M.; Tan, T. T. Y., Nanostructure control of graphene-composited TiO_2 by a one-step solvothermal approach for high performance dye-sensitized solar cells. *Nanoscale* **2011**, *3*, 4613–4616.
52. Arsyad, W. S.; Pristianti, A.; Agusu, L.; Hidayat, R., rGO based photo-anode in dye-sensitized solar cells (DSSC) and its photovoltaic characteristics. *IOP Conference Series: Materials Science and Engineering* **2019**, *622*, 012008.
53. Wojcik, A.; Kamat, P. V., Reduced graphene oxide and porphyrin: An interactive affair in 2-D. *ACS Nano* **2010**, *4*, 6697–6706.
54. Liang, J.; Zhang, G.; Yang, J.; Sun, W.; Shi, M., TiO_2 hierarchical nanostructures: Hydrothermal fabrication and application in dye-sensitized solar cells. *AIP Advances* **2015**, *5*, 017141.
55. Nouri, E.; Mohammadi, M. R.; Lianos, P., Impact of preparation method of TiO_2-RGO nanocomposite photoanodes on the performance of dye-sensitized solar cells. *Electrochimica Acta* **2016**, *219*, 38–48.
56. Mehmood, U.; Ahmad, S. H. A.; Khan, A. U. H.; Qaiser, A. A., Co-sensitization of graphene/TiO_2 nanocomposite thin films with

ruthenizer and metal free organic photosensitizers for improving the power conversion efficiency of dye-sensitized solar cells (DSSCs). *Solar Energy* **2018**, *170*, 47–55.

57. Salam, Z.; Vijayakumar, E.; Subramania, A.; Sivasankar, N.; Mallick, S., Graphene quantum dots decorated electrospun TiO_2 nanofibers as an effective photoanode for dye sensitized solar cells. *Solar Energy Materials and Solar Cells* **2015**, *143*, 250–259.
58. Xiang, Z.; Zhou, X.; Wan, G.; Zhang, G.; Cao, D., Improving energy conversion efficiency of dye-sensitized solar cells by modifying TiO_2 photoanodes with nitrogen-reduced graphene oxide. *ACS Sustainable Chemistry & Engineering* **2014**, *2*, 1234–1240.
59. Jayabal, P.; Gayathri, S.; Sasirekha, V.; Mayandi, J.; Ramakrishnan, V., Preparation and characterization of ZnO/graphene nanocomposite for improved photovoltaic performance. *Journal of Nanoparticle Research* **2014**, *16*, 2640.
60. Xu, F.; Chen, J.; Wu, X.; Zhang, Y.; Wang, Y.; Sun, J.; Bi, H.; Lei, W.; Ni, Y.; Sun, L., Graphene scaffolds enhanced photogenerated electron transport in ZnO photoanodes for high-efficiency dye-sensitized solar cells. *The Journal of Physical Chemistry C* **2013**, *117*, 8619–8627.
61. Sasikumar, R.; Chen, T.-W.; Chen, S.-M.; Rwei, S.-P.; Ramaraj, S. K., Developing the photovoltaic performance of dye-sensitized solar cells (DSSCs) using a SnO_2-doped graphene oxide hybrid nano-composite as a photo-anode. *Optical Materials* **2018**, *79*, 345–352.
62. Nemala, S. S.; Kartikay, P.; Prathapani, S.; Bohm, H. L. M.; Bhargava, P.; Bohm, S.; Mallick, S., Liquid phase high shear exfoliated graphene nanoplatelets as counter electrode material for dye-sensitized solar cells. *Journal of Colloid and Interface Science* **2017**, *499*, 9–16.
63. Hong, W.; Xu, Y.; Lu, G.; Li, C.; Shi, G., Transparent graphene/PEDOT–PSS composite films as counter electrodes of dye-sensitized solar cells. *Electrochemistry Communications* **2008**, *10*, 1555–1558.
64. Huh, S. H.; Choi, S.-H.; Ju, H.-M.; Song, C. K.; Park, S. H.; Kim, B. M., A catalytic graphene oxide film for a dye-sensitized solar cell. *Journal of the Korean Physical Society* **2010**, *57*, 1653–1656.
65. Hasin, P.; Alpuche-Aviles, M. A.; Wu, Y., Electrocatalytic activity of graphene multilayers toward I^-/I_3^-: Effect of preparation conditions and polyelectrolyte modification. *The Journal of Physical Chemistry C* **2010**, *114*, 15857–15861.
66. Roy-Mayhew, J. D.; Bozym, D. J.; Punckt, C.; Aksay, I. A., Functionalized graphene as a catalytic counter electrode in dye-sensitized solar cells. *ACS Nano* **2010**, *4*, 6203–6211.

67. Kavan, L.; Yum, J. H.; Grätzel, M., Optically transparent cathode for dye-sensitized solar cells based on graphene nanoplatelets. *ACS Nano* **2011**, *5*, 165–172.

68. Kavan, L.; Yum, J.-H.; Nazeeruddin, M. K.; Grätzel, M., Graphene nanoplatelet cathode for Co(III)/(II) mediated dye-sensitized solar cells. *ACS Nano* **2011**, *5*, 9171–9178.

69. Nagavolu, C.; Susmitha, K.; Raghavender, M.; Giribabu, L.; Bhanu Sankara Rao, K.; Smith, C. T. G.; Mills, C. A.; Silva, S. R. P.; Srikanth, V. V. S. S., Pt-free spray coated reduced graphene oxide counter electrodes for dye sensitized solar cells. *Solar Energy* **2016**, *137*, 143–147.

70. Xue, Y.; Liu, J.; Chen, H.; Wang, R.; Li, D.; Qu, J.; Dai, L., Nitrogen-doped graphene foams as metal-free counter electrodes in high-performance dye-sensitized solar cells. *Angewandte Chemie International Edition* **2012**, *51*, 12124–12127.

71. Wei, W.; Sun, K.; Hu, Y. H., Synthesis of 3D cauliflower-fungus-like graphene from CO2 as a highly efficient counter electrode material for dye-sensitized solar cells. *Journal of Materials Chemistry A* **2014**, *2*, 16842–16846.

72. Ma, J.; Shen, W.; Li, C.; Zheng, J.; Yu, F., Graphene cryogel-based counter electrode materials freeze-dried using different solution media for dye-sensitized solar cells. *Chemical Engineering Journal* **2017**, *319*, 155–162.

73. Wu, W.-T.; Yang, S.-H.; Hsu, C.-M.; Wu, W.-T., Study of graphene nanoflake as counter electrode in dye sensitized solar cells. *Diamond and Related Materials* **2016**, *65*, 91–95.

74. Pan, D.; Feng, C.; Wang, L.; Liu, Y.; Chen, Z.; Shi, W.; Li, Z., Binder-free graphene organogels as cost-efficient counter electrodes for dye-sensitized solar cells. *Electrochimica Acta* **2016**, *191*, 946–953.

75. Yao, J.; Zhang, K.; Wang, W.; Zuo, X.; Yang, Q.; Tang, H.; Wu, M.; Li, G., Remarkable enhancement in the photoelectric performance of uniform flower-like mesoporous Fe3O4 wrapped in nitrogen-doped graphene networks. *ACS Applied Materials & Interfaces* **2018**, *10*, 19564–19572.

76. Meng, X.; Yu, C.; Song, X.; Iocozzia, J.; Hong, J.; Rager, M.; Jin, H.; Wang, S.; Huang, L.; Qiu, J.; Lin, Z., Scrutinizing defects and defect density of selenium-doped graphene for high-efficiency triiodide reduction in dye-sensitized solar cells. *Angewandte Chemie (International ed. in English)* **2018**, *57*, 4682–4686.

77. Cui, X.; Xiao, J.; Wu, Y.; Du, P.; Si, R.; Yang, H.; Tian, H.; Li, J.; Zhang, W.-H.; Deng, D.; Bao, X., A graphene composite material with single cobalt active sites: A highly efficient counter electrode for dye-sensitized solar cells. *Angewandte Chemie International Edition* **2016**, *55*, 6708–6712.
78. Jia, Y.; Shi, Y.; Qiu, J.; Hao, C., Theoretical design and experimental synthesis of counter electrode for dye-sensitized solar cells: Amino-functionalized graphene. *Journal of Energy Chemistry* **2016**, *25*, 861–867.
79. Zhai, P.; Wei, T.-C.; Chang, Y.-H.; Huang, Y.-T.; Yeh, W.-T.; Su, H.; Feng, S.-P., High electrocatalytic and wettable nitrogen-doped microwave-exfoliated graphene nanosheets as counter electrode for dye-sensitized solar cells. *Small* **2014**, *10*, 3347–3353.
80. Wu, M.; Sun, M.; Zhou, H.; Ma, J.-Y.; Ma, T., Carbon counter electrodes in dye-sensitized and perovskite solar cells. *Advanced Functional Materials* **2020**, *30*, 1906451.
81. Yuan, H.; Jiao, Q.; Liu, J.; Liu, X.; Yang, H.; Zhao, Y.; Wu, Q.; Shi, D.; Li, H., Ultrathin-walled Co_9S_8 nanotube/reduced graphene oxide composite as an efficient electrocatalyst for the reduction of triiodide. *Journal of Power Sources* **2016**, *336*, 132–142.
82. Yuan, H.; Liu, J.; Jiao, Q.; Li, Y.; Liu, X.; Shi, D.; Wu, Q.; Zhao, Y.; Li, H., Sandwich-like octahedral cobalt disulfide/reduced graphene oxide as an efficient Pt-free electrocatalyst for high-performance dye-sensitized solar cells. *Carbon* **2017**, *119*, 225–234.
83. Yuan, H.; Liu, J.; Li, H.; Su, K.; Liu, X.; Li, Y.; Shi, D.; Wu, Q.; Zhao, Y.; Jiao, Q., Rational integration of hierarchical structural $CoS_{1.097}$ nanosheets/reduced graphene oxide nanocomposites with enhanced electrocatalytic performance for triiodide reduction. *Carbon* **2018**, *126*, 514–521.
84. Lin, J.-Y.; Chan, C.-Y.; Chou, S.-W., Electrophoretic deposition of transparent MoS_2–graphene nanosheet composite films as counter electrodes in dye-sensitized solar cells. *Chemical Communications* **2013**, *49*, 1440–1442.
85. Lin, J. Y.; Su, A. L.; Chang, C. Y.; Hung, K. C.; Lin, T. W., Molybdenum disulfide/reduced graphene oxide-carbon nanotube hybrids as efficient catalytic materials in dye-sensitized solar cells. *ChemElectroChem* **2015**, *2*, 720–725.
86. Zhang, X.; Yang, Y.; Guo, S.; Hu, F.; Liu, L., Mesoporous $Ni_{0.85}Se$ nanospheres grown in situ on graphene with high performance in

dye-sensitized solar cells. *ACS Applied Materials & Interfaces* **2015**, *7*, 8457–8464.

87. Lin, J.-Y.; Su, A.-L.; Chang, C.-Y.; Hung, K.-C.; Lin, T.-W., Molybdenum disulfide/reduced graphene oxide–carbon nanotube hybrids as efficient catalytic materials in dye-sensitized solar cells. *ChemElectroChem* **2015**, *2*, 720–725.

88. Khoo, S. Y.; Miao, J.; Yang, H. B.; He, Z.; Leong, K. C.; Liu, B.; Tan, T. T. Y., One-step hydrothermal tailoring of $NiCo_2S_4$ nanostructures on conducting oxide substrates as an efficient counter electrode in dye-sensitized solar cells. *Advanced Materials Interfaces* **2015**, *2*, 1500384.

89. Krishnapriya, R.; Praneetha, S.; Rabel, A. M.; Vadivel Murugan, A., Energy efficient, one-step microwave-solvothermal synthesis of a highly electro-catalytic thiospinel $NiCo_2S_4$/graphene nanohybrid as a novel sustainable counter electrode material for Pt-free dye-sensitized solar cells. *Journal of Materials Chemistry C* **2017**, *5*, 3146–3155.

90. Marchezi, P. E.; Sonai, G. G.; Hirata, M. K.; Schiavon, M. A.; Nogueira, A. F., Understanding the role of reduced graphene oxide in the electrolyte of dye-sensitized solar cells. *The Journal of Physical Chemistry C* **2016**, *120*, 23368–23376.

91. Velten, J. A.; Carretero-González, J.; Castillo-Martínez, E.; Bykova, J.; Cook, A.; Baughman, R.; Zakhidov, A., Photoinduced optical transparency in dye-sensitized solar cells containing graphene nanoribbons. *The Journal of Physical Chemistry C* **2011**, *115*, 25125–25131.

92. Ahmad, I.; Khan, U.; Gun'ko, Y. K., Graphene, carbon nanotube and ionic liquid mixtures: Towards new quasi-solid state electrolytes for dye sensitised solar cells. *Journal of Materials Chemistry* **2011**, *21*, 16990–16996.

93. Jung, M.-H.; Kang, M. G.; Chu, M.-J., Iodide-functionalized graphene electrolyte for highly efficient dye-sensitized solar cells. *Journal of Materials Chemistry* **2012**, *22*, 16477–16483.

94. Brennan, L. J.; Barwich, S. T.; Satti, A.; Faure, A.; Gun'ko, Y. K., Graphene–ionic liquid electrolytes for dye sensitised solar cells. *Journal of Materials Chemistry A* **2013**, *1*, 8379–8384.

95. Zheng, J., Graphene tailored polymer gel electrolytes for 9.1%-efficiency quasi-solid-state dye-sensitized solar cells. *Journal of Power Sources* **2017**, *348*, 239–245.

96. Lin, B.; Shang, H.; Chu, F.; Ren, Y.; Yuan, N.; Jia, B.; Zhang, S.; Yu, X.; Wei, Y.; Ding, J., Ionic liquid-tethered graphene oxide/ionic liquid electrolytes for highly efficient dye sensitized solar cells. *Electrochimica Acta* **2014**, *134*, 209–214.

97. Chan, Y.-F.; Wang, C.-C.; Chen, C.-Y., Quasi-solid DSSC based on a gel-state electrolyte of PAN with 2-D graphenes incorporated. *Journal of Materials Chemistry A* **2013**, *1*, 5479–5486.

98. Yuan, S.; Tang, Q.; Hu, B.; Ma, C.; Duan, J.; He, B., Efficient quasi-solid-state dye-sensitized solar cells from graphene incorporated conducting gel electrolytes. *Journal of Materials Chemistry A* **2014**, *2*, 2814–2821.

99. Yuan, S.; Tang, Q.; He, B.; Zhao, Y., Multifunctional graphene incorporated conducting gel electrolytes in enhancing photovoltaic performances of quasi-solid-state dye-sensitized solar cells. *Journal of Power Sources* **2014**, *260*, 225–232.

100. Nath, B. C.; Das, D.; Kamrupi, I. R.; Mohan, K. J.; Ahmed, G. A.; Dolui, S. K., An efficient quasi solid state dye sensitized solar cell based on polyethylene glycol/graphene nanosheet gel electrolytes. *RSC Advances* **2015**, *5*, 95385–95393.

101. Chen, X.; Tang, Q.; He, B.; Chen, H., Graphene-incorporated quasi-solid-state dye-sensitized solar cells. *RSC Advances* **2015**, *5*, 43402–43407.

102. Shrivatsav, R.; Mahalingam, V.; Narayanan, E. R. L.; Balaji, N. N.; Balu, M.; Prasad, R. K.; Kumaresan, D., Characterization of poly methyl methaacrylate and reduced graphene oxide composite for application as electrolyte in dye sensitized solar cells. *Materials Research Express* **2018**, *5*, 046204.

103. Li, X.; Zhou, Y.; Chen, J.; Yang, J.; Zheng, Z.; Wu, W.; Hua, J.; Tian, H., Stacked graphene platelet nanofibers dispersed in the liquid electrolyte of highly efficient cobalt-mediator-based dye-sensitized solar cells. *Chemical Communications* **2015**, *51*, 10349–10352.

104. Venkatesan, S.; Surya Darlim, E.; Tsai, M.-H.; Teng, H.; Lee, Y.-L., Graphene oxide sponge as nanofillers in printable electrolytes in high-performance quasi-solid-state dye-sensitized solar cells. *ACS Applied Materials & Interfaces* **2018**, *10*, 10955–10964.

105. Venkatesan, S.; Darlim, E. S.; Liu, I. P.; Lee, Y.-L., Performance enhancement effects of dispersed graphene oxide sponge nanofillers on the liquid electrolytes of dye-sensitized solar cells. *Carbon* **2018**, *132*, 71–77.

106. Yan, X.; Cui, X.; Li, B.; Li, L.-s., Large, solution-processable graphene quantum dots as light absorbers for photovoltaics. *Nano Letters* **2010**, *10*, 1869–1873.

107. Williams, K. J.; Nelson, C. A.; Yan, X.; Li, L.-S.; Zhu, X., Hot electron injection from graphene quantum dots to TiO_2. *ACS Nano* **2013**, *7*, 1388–1394.

108. Mihalache, I.; Radoi, A.; Mihaila, M.; Munteanu, C.; Marin, A.; Danila, M.; Kusko, M.; Kusko, C., Charge and energy transfer interplay in hybrid sensitized solar cells mediated by graphene quantum dots. *Electrochimica Acta* **2015**, *153*, 306–315.

109. San Esteban, A. C. M.; Enriquez, E. P., Graphene–anthocyanin mixture as photosensitizer for dye-sensitized solar cell. *Solar Energy* **2013**, *98*, 392–399.

110. Gatti, T.; Manfredi, N.; Boldrini, C.; Lamberti, F.; Abbotto, A.; Menna, E., A D-π-A organic dye: Reduced graphene oxide covalent dyad as a new concept photosensitizer for light harvesting applications. *Carbon* **2017**, *115*, 746–753.

111. Niu, G.; Guo, X.; Wang, L., Review of recent progress in chemical stability of perovskite solar cells. *Journal of Materials Chemistry A* **2015**, *3*, 8970–8980.

112. Zheng, S.; Wang, G.; Liu, T.; Lou, L.; Xiao, S.; Yang, S., Materials and structures for the electron transport layer of efficient and stable perovskite solar cells. *Science China Chemistry* **2019**, *62*, 800–809.

113. Wang, K.; Olthof, S.; Subhani, W. S.; Jiang, X.; Cao, Y.; Duan, L.; Wang, H.; Du, M.; Liu, S., Novel inorganic electron transport layers for planar perovskite solar cells: Progress and prospective. *Nano Energy* **2020**, *68*, 104289.

114. Mohamad Noh, M. F.; Teh, C. H.; Daik, R.; Lim, E. L.; Yap, C. C.; Ibrahim, M. A.; Ahmad Ludin, N.; Mohd Yusoff, A. R. b.; Jang, J.; Mat Teridi, M. A., The architecture of the electron transport layer for a perovskite solar cell. *Journal of Materials Chemistry C* **2018**, *6*, 682–712.

115. Mahmood, K.; Sarwar, S.; Mehran, M. T., Current status of electron transport layers in perovskite solar cells: Materials and properties. *RSC Advances* **2017**, *7*, 17044–17062.

116. Wang, J. T.-W.; Ball, J. M.; Barea, E. M.; Abate, A.; Alexander-Webber, J. A.; Huang, J.; Saliba, M.; Mora-Sero, I.; Bisquert, J.; Snaith, H. J.; Nicholas, R. J., Low-temperature processed electron collection layers of graphene/TiO_2 nanocomposites in thin film perovskite solar cells. *Nano Letters* **2014**, *14*, 724–730.

117. Han, G. S.; Song, Y. H.; Jin, Y. U.; Lee, J.-W.; Park, N.-G.; Kang, B. K.; Lee, J.-K.; Cho, I. S.; Yoon, D. H.; Jung, H. S., Reduced graphene oxide/mesoporous TiO_2 nanocomposite-based perovskite solar cells. *ACS Applied Materials & Interfaces* **2015**, *7*, 23521–23526.

118. Agresti, A.; Pescetelli, S.; Palma, A. L.; Del Rio Castillo, A. E.; Konios, D.; Kakavelakis, G.; Razza, S.; Cinà, L.; Kymakis, E.; Bonaccorso, F.; Di Carlo, A., Graphene interface engineering for perovskite solar modules: 12.6% power conversion efficiency over 50 cm^2 active area. *ACS Energy Letters* **2017**, *2*, 279–287.

119. O'Keeffe, P.; Catone, D.; Paladini, A.; Toschi, F.; Turchini, S.; Avaldi, L.; Martelli, F.; Agresti, A.; Pescetelli, S.; Del Rio Castillo, A. E.; Bonaccorso, F.; Di Carlo, A., Graphene-induced improvements of perovskite solar cell stability: Effects on hot-carriers. *Nano Letters* **2019**, *19*, 684–691.

120. Agresti, A.; Pescetelli, S.; Cinà, L.; Konios, D.; Kakavelakis, G.; Kymakis, E.; Carlo, A. D., Efficiency and stability enhancement in perovskite solar cells by inserting lithium-neutralized graphene oxide as electron transporting layer. *Advanced Functional Materials* **2016**, *26*, 2686–2694.

121. Biccari, F.; Gabelloni, F.; Burzi, E.; Gurioli, M.; Pescetelli, S.; Agresti, A.; Del Rio Castillo, A. E.; Ansaldo, A.; Kymakis, E.; Bonaccorso, F.; Di Carlo, A.; Vinattieri, A., Graphene-based electron transport layers in perovskite solar cells: A step-up for an efficient carrier collection. *Advanced Energy Materials* **2017**, *7*, 1701349.

122. Tavakoli, M. M.; Tavakoli, R.; Nourbakhsh, Z.; Waleed, A.; Virk, U. S.; Fan, Z., High efficiency and stable perovskite solar cell using ZnO/rGO QDs as an electron transfer layer. *Advanced Materials Interfaces* **2016**, *3*, 1500790.

123. Zhao, X.; Tao, L.; Li, H.; Huang, W.; Sun, P.; Liu, J.; Liu, S.; Sun, Q.; Cui, Z.; Sun, L.; Shen, Y.; Yang, Y.; Wang, M., Efficient planar perovskite solar cells with improved fill factor via interface engineering with graphene. *Nano Letters* **2018**, *18*, 2442–2449.

124. Zhu, M.; Liu, W.; Ke, W.; Xie, L.; Dong, P.; Hao, F., Graphene-modified tin dioxide for efficient planar perovskite solar cells with enhanced electron extraction and reduced hysteresis. *ACS Applied Materials & Interfaces* **2019**, *11*, 666–673.

125. Kakavelakis, G.; Maksudov, T.; Konios, D.; Paradisanos, I.; Kioseoglou, G.; Stratakis, E.; Kymakis, E., Efficient and highly air stable planar inverted perovskite solar cells with reduced graphene oxide doped

PCBM electron transporting layer. *Advanced Energy Materials* **2017**, *7*, 1602120.

126. Nouri, E.; Mohammadi, M. R.; Lianos, P., Inverted perovskite solar cells based on lithium-functionalized graphene oxide as an electron-transporting layer. *Chemical Communications* **2017**, *53*, 1630–1633.

127. Nouri, E.; Mohammadi, M. R.; Lianos, P., Improving the stability of inverted perovskite solar cells under ambient conditions with graphene-based inorganic charge transporting layers. *Carbon* **2018**, *126*, 208–214.

128. Kim, G.-H.; Jang, H.; Yoon, Y. J.; Jeong, J.; Park, S. Y.; Walker, B.; Jeon, I.-Y.; Jo, Y.; Yoon, H.; Kim, M.; Baek, J.-B.; Kim, D. S.; Kim, J. Y., Fluorine functionalized graphene nano platelets for highly stable inverted perovskite solar cells. *Nano Letters* **2017**, *17*, 6385–6390.

129. Bakr, Z. H.; Wali, Q.; Fakharuddin, A.; Schmidt-Mende, L.; Brown, T. M.; Jose, R., Advances in hole transport materials engineering for stable and efficient perovskite solar cells. *Nano Energy* **2017**, *34*, 271–305.

130. Luo, Q.; Zhang, Y.; Liu, C.; Li, J.; Wang, N.; Lin, H., Iodide-reduced graphene oxide with dopant-free spiro-OMeTAD for ambient stable and high-efficiency perovskite solar cells. *Journal of Materials Chemistry A* **2015**, *3*, 15996–16004.

131. Wen, X.; Wu, J.; Gao, D.; Lin, C., Interfacial engineering with amino-functionalized graphene for efficient perovskite solar cells. *Journal of Materials Chemistry A* **2016**, *4*, 13482–13487.

132. Cao, J.; Liu, Y.-M.; Jing, X.; Yin, J.; Li, J.; Xu, B.; Tan, Y.-Z.; Zheng, N., Well-defined thiolated nanographene as hole-transporting material for efficient and stable perovskite solar cells. *Journal of the American Chemical Society* **2015**, *137*, 10914–10917.

133. Palma, A. L.; Cinà, L.; Pescetelli, S.; Agresti, A.; Raggio, M.; Paolesse, R.; Bonaccorso, F.; Di Carlo, A., Reduced graphene oxide as efficient and stable hole transporting material in mesoscopic perovskite solar cells. *Nano Energy* **2016**, *22*, 349–360.

134. Lee, D.-Y.; Na, S.-I.; Kim, S.-S., Graphene oxide/PEDOT:PSS composite hole transport layer for efficient and stable planar heterojunction perovskite solar cells. *Nanoscale* **2016**, *8*, 1513–1522.

135. Kim, J. M.; Jang, C. W.; Kim, J. H.; Kim, S.; Choi, S.-H., Use of $AuCl_3$-doped graphene as a protecting layer for enhancing the stabilities of inverted perovskite solar cells. *Applied Surface Science* **2018**, *455*, 1131–1136.

136. Feng, S.; Yang, Y.; Li, M.; Wang, J.; Cheng, Z.; Li, J.; Ji, G.; Yin, G.; Song, F.; Wang, Z.; Li, J.; Gao, X., High-performance perovskite solar cells engineered by an ammonia modified graphene oxide interfacial layer. *ACS Applied Materials & Interfaces* **2016**, *8*, 14503–14512.

137. Wu, Z.; Bai, S.; Xiang, J.; Yuan, Z.; Yang, Y.; Cui, W.; Gao, X.; Liu, Z.; Jin, Y.; Sun, B., Efficient planar heterojunction perovskite solar cells employing graphene oxide as hole conductor. *Nanoscale* **2014**, *6*, 10505–10510.

138. Li, W.; Dong, H.; Guo, X.; Li, N.; Li, J.; Niu, G.; Wang, L., Graphene oxide as dual functional interface modifier for improving wettability and retarding recombination in hybrid perovskite solar cells. *Journal of Materials Chemistry A* **2014**, *2*, 20105–20111.

Chapter 2

Boron Nitride-Based Electrochemical Energy Storage Devices

Shanmugaraj Pavithra,[a] Ayyasamy Sakunthala,[a] Bhojan Vidhya,[a] Swaminathan Rajesh,[a] Raju Nandhakumar,[a] Mogalahalli Venkatesh Reddy,[b] Karim Zaghib,[c] and Rajan Jose[d]

[a]*Division of Physical Sciences, Karunya Institute of Technology and Sciences, Coimbatore 641 114, Tamil Nadu, India*
[b]*Nouveau Monde Graphite, 481 Rue Brassard, Saint-Michel-de-Saints, QC J0K 3B0, Canada*
[c]*Department of Chemical and Materials Engineering, Concordia University, 1455 Boulevard de Maisonneuve, Montréal, QC H3G 1M8, Canada*
[d]*Nanostructured Renewable Energy Materials Laboratory, Faculty of Industrial Sciences & Technology, University Malaysia Pahang Al-Sultan Abdullah, 26300 Kuantan, Malaysia*

sakunthala@karunya.edu, meetsaku@gmail.com

2.1 Introduction

In recent years, two-dimensional (2D) materials have gained high attention among researchers for various applications [1–4]. Among them, boron nitride (BN), a manmade 2D material, has become a hot material of research interest. In 1842, Balmain synthesized boron nitride using molten boric acid and potassium

Layered Materials: Synthesis and Applications in Energy and Environmental Remediation
Edited by Alagarsamy Pandikumar, Kandasamy Jothivenkatachalam, and Subramanian Moscow

ISBN 978-981-5129-17-5 (Hardcover), 978-1-003-50831-1 (eBook)
www.jennystanford.com

cyanide [5]. However, the product was unstable and it took hundreds of years of research to form a stable powder of boron nitride. The fabrication of BN greatly developed only during the past century especially when hot pressing of BN was realized after the 1960s [6]. BN is an isoelectronic and isomorphic to carbon in any lattice structure, from zero-dimensional (0D) cage, one-dimensional (1D) nanotube, two-dimensional (2D) monolayer to three-dimensional (3D) diamond-like crystal structures [6, 7]. Jiang et al. [8] have discussed in detail the structure, synthesis routes, properties, and applications of one-dimensional (1D) nanotubes and nanoribbons, 2D nanospheres and nanosheets, and three-dimensional (3D) nanoporous materials of boron nitride. Figure 2.1 shows the structural model of 2D, 1D, and 0D BN nanostructures.

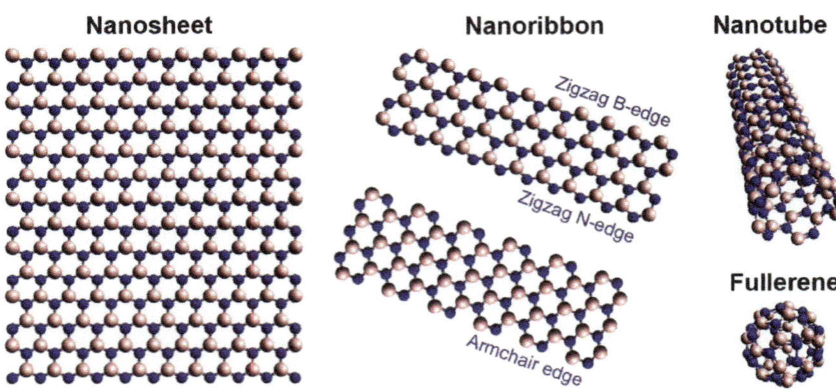

Figure 2.1 Structural model of 2D, 1D, and 0D BN nanostructures. The edge of a nanosheet or nanoribbon can be either zigzag (B- or N-edged) or armchair (BN pair-edged) [22].

Boron nitride is a typical III–V group compound. Equal amounts of B and N atoms are arranged in the direction of ABAB in the C axis making up the structure of BN with the bond length 1.45 Å and lattice parameters $a = b = 0.2504$ nm. BN has a layered structure similar to graphene. Each layer is connected through weak van der Waals forces with a layer spacing of 0.6661 nm [9]. Unlike graphene, BN is an insulator with a high band gap (E_g = 4–6 eV) [10]. It has high chemical stability, thermal conductivity, electrical resistivity, thermal shock resistance, melting point (>3000 K), low dielectric constant, and is non-toxic in nature [11, 12]. These properties make BN a promising material for applications in batteries [13], supercapacitors [14], gas sensing [15], coating layers [16], optoelectronics [17], radiation detectors [18], fuel cells [19], drug delivery [20], biomedical [21], etc.

In general, the synthesis routes of a material can be classified into two categories: top-down and bottom-up routes, based on the dimensions, purity, cost, and time required to obtain end products [23]. Top-down methods, including ball-milling [24], high-pressure micro fluidization [25], and exfoliation techniques such as mechanical exfoliation [26], liquid exfoliation [27], chemical-assisted exfoliation [28], sonication-assisted exfoliation [29], and acid exfoliation [30] are used to synthesize BN. Bottom-up methods for BN synthesis including chemical vapor deposition (CVD) [31], thermal vapor deposition (TCVD) [32], controlled synthesis using atmospheric pressure CVD (APCVD) [33], pulsed laser deposition [34], magnetron sputtering [35], molecular beam epitaxy [36], and non-covalent [37], ionic [38], and covalent [39] functionalization are the approaches to completely functionalize the BN nanosheets (NS). Synthesis methods and parameters including precursors, temperature, and pressure greatly influence the crystal and morphological structures of 2D BN [40].

Boron nitride has four major polymorphs: (a) cubic boron nitride (c-BN) that has the same crystal lattice as diamond, (b) wurtzitic boron nitride (w-BN) that has the same crystal lattice as lonsdaleite, (c) hexagonal boron nitride (h-BN) also known as "white graphene" that has the same crystal lattice as graphite, and (d) rhombohedral boron nitride (r-BN) that has the same crystal lattice as rhombohedral graphite [10, 41]. These polymorphs have different structural and physical properties. Figure 2.2 shows crystal structures of (a) cubic, (b) wurtzite, (c) hexagonal, and (d) rhombohedral BN. Some other polymorphs are amorphous BN (a-BN), turbostratic BN (t-BN), and rocksalt BN (rs-BN). The polymorph hexagonal BN is studied widely because of its excellent chemical and thermal stability than other polymorphs. A single layer of h-BN is called BN nanosheets (BNNS) [42].

In this chapter, a detailed discussion has been made on BN for its application as an electrode in popular electrochemical energy storage devices, viz. batteries and supercapacitors. The synthesis of electrodes (cathode and anode), electrolytes, and separators using BN and their electrical and electrochemical properties are explained in detail in this chapter.

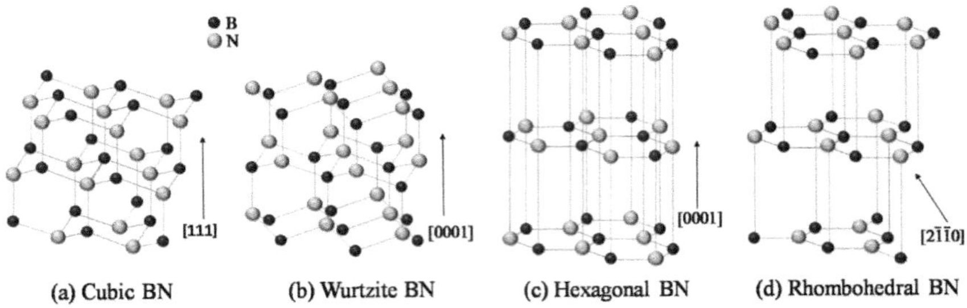

Figure 2.2 Crystal structures of (a) cubic, (b) wurtzite, (c) hexagonal, and (d) rhombohedral BN [10].

2.1.1 Highlights of Boron Nitride as a Cathode and Anode Electrode Material in Batteries

The review described in this section highlights the efficacy of BN in the electrode or electrode additive, and in polymer separators, for the present generation lithium-ion batteries, lithium metal batteries, and for the batteries at research levels like sulfur batteries, magnesium batteries, aluminum batteries, sodium-ion batteries, etc. One of the primary cost factors of the current lithium-ion batteries is the electrode material containing cobalt; therefore, the battery industry strives for new materials and alternative technologies for a cost-effective solution when compared to the present-day batteries [43].

2.2 BN in Cathode Materials for Batteries

2.2.1 BN as a Sulfur Fixer

Aluminum-sulfur batteries (ASBs) were found to have better energy density than aluminum-ion batteries made with carbon-based or other composite cathodes [44–47]. However, the serious problem is the dissolution of polysulfide compounds as the discharge product, which limits its life span. Here comes the importance of BN as a solution for this impediment, which can act as a fixer of sulfur and polysulfide, as reported by Zhang et al. [48]. The research group has used the combination of ball-milled BN/S/C as a cathode material and demonstrated a discharge capacity of up to 532 mAh g^{-1} at a current density of 100 mA g^{-1}, when compared with the current state-of-the-art cathode material for AIB's. Any of the 2D materials will have more active sites on their surface, which is useful for the adsorption of elemental sulfur and polysulfide compounds and could act as an S fixer [48]. Interestingly, on comparing, S-containing transition metal sulfides (MoS_2 and WS_2) and S-free 2D (BN) layered materials by decorating them with S and C by ball milling, BN proved itself as a good fixer of S and stored capacity of S by bonding with S and/or sulfide compounds (Fig. 2.3). Uniform distribution of S on BN and also the formation of sulfide compounds were confirmed by the disappearance of elemental S diffraction peak in the cathode after a cycling test. Thus, the above group concluded that boron nitride is a promising material for the adsorption of polysulfides, which leads to a long-life span with superior electrochemical performance of ASB (Table 2.1). Aluminum is one of the most common elements on the earth's crust that has high safety and a higher reduction potential (−1.76 V versus a SHE). The above battery has high theoretical gravimetric and volumetric capacities of 2978 mAh g^{-1} and 8034 mAh cm^{-3}, respectively. Hence, fixing the problems associated with the dissolution of discharge products with the potential properties of BN is to be appreciated.

2.2.2 BN as Efficient Cathode for Lithium-Sulfur Battery

Lithium-sulfur (Li-S) batteries are leading the technological advancement in the battery industry. There are only scarce works reported on boron nitride for Li-S applications. When the 2D graphene is supported with BN nanosheets, the synergetic effects between them were found to enhance the adsorption of the polysulfide in a wide range [49], in the case of Li-S batteries. The stronger adsorption of polysulfides is enhanced by BN than graphene, which was proved by the electrochemical deposition of Li_2S, which showed that most of the discharged products were deposited at the edges of the BN nanosheets. Also, the electrode gives a specific capacity of 1553 mAh g^{-1}, which is equivalent to 93% of the theoretical capacity of sulfur.

Table 2.1 Electrochemical performance of the BN as a cathode material for batteries

Cathode composition	Method of preparation	Device	Voltage range (V)	Current density	Capacity mAh g^{-1}	Cycle	Capacity mAh g^{-1}	Ref.
v-BN	Solution calcining	Li-S	1.7–2.6	0.1 C	1262	150	770	[13]
BN/S/C	Ball-milling	ASB	0.05–2.2	100 mA g^{-1}	532	300	300	[48]
BN/graphene-S	Classical melt-diffusion	Li-S	1.7–2.7	0.1 C	1553	30	1400	[49]
F-h-BNNS	Chemical process	Mg	0.0–3.5	15 mA g^{-1}	50	80	–	[50]
G/BN-20%	Vacuum filtration	Mg	0.0–2.6	25 mA g^{-1}	105	80	105	[51]
BN-MnO_2	Disk-like pellet	Alkaline	0.0–2.0	0.5 mA cm^2	175	25	110	[52]
Co-BN	Ball-milling	LIB	0.60–1.05	50 mA g^{-1}	350	50	250	[53]

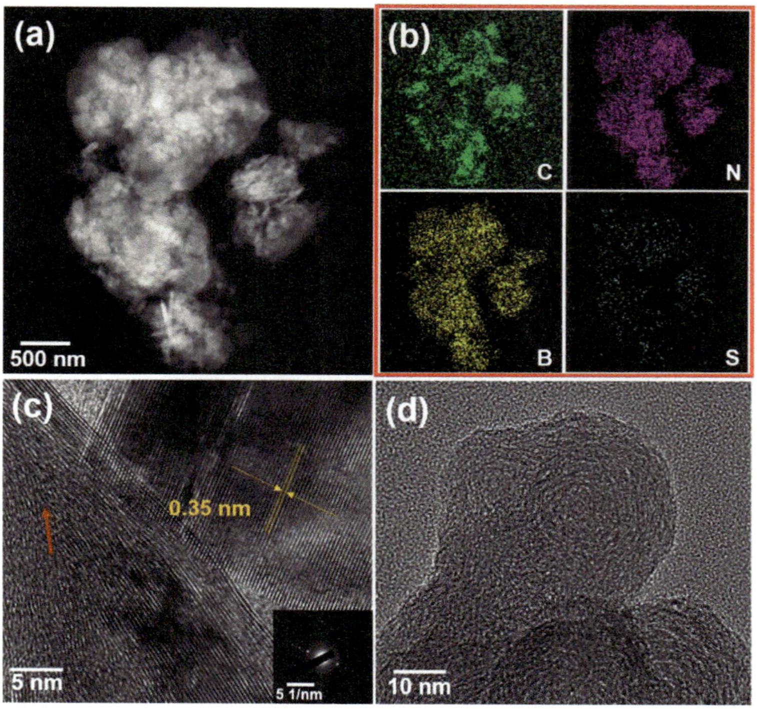

Figure 2.3 (a) TEM image, (b) EDX mapping, (c) HRTEM image of BN, and (d) HRTEM image of C in the BN/S/C material [48].

Similarly, cathodes of Li-S batteries were constructed using a few layers of BN with engineered nitrogen vacancies (v-BN) [13]. The positive vacancy defects existing in BN nanosheets because of the unsaturated B atoms and lots of N vacancy defects with an N/B ratio of 0.708 enhance the interaction with lithium-sulfur compounds, lithium-ion diffusion, and electro-catalytic activity for their conversion. When compared with pristine BN, v-BN shows the superior adsorptive ability of polysulfides, the kinetics of polysulfide conversion, large diffusion coefficient, and electrochemical performance (Table 2.1).

2.2.3 Synergistic Effect of Graphene/BN for Magnesium Batteries

Magnesium (Mg) batteries, owing to their high safety and lower cost compared to LIBs, have attained appreciable research interest.

However, less reversible Mg insertion/extraction in the cathode is observed because of the strongly polarizing divalent Mg ions [50, 51]. A good Mg insertion/extraction was obtained when graphene-supported boron nitride nanosheet (G/BN) composites were developed with 20, 30, and 40% of BN as cathode materials [51]. The composite G/BN-20% showed low charge transfer resistance (R_{ct}) with good reversibility, high current density, and high discharge capacity (Table 2.1). This was mainly attributed to the proper ratio of graphene and BN composites, which enabled a favorable synergistic effect [51]. Fluorinated hexagonal boron nitride nanosheets (F-h-BNNSs) were used as active materials for cathodes. Due to the insulating nature of h-BNNS, doping highly electronegative fluorine in the vulnerable sites of h-BNNS minimizes the band gap. The 'F' doping mechanism into the layered h-BNNS is represented by Eq. (2.1) [50].

$$\text{BNNSs} + 2\text{HBF}_4 - \text{H}_2\text{O} \leftrightarrow 2\text{F}_{0.6}\text{BNNSs} + 2\text{H}_3\text{O}^+ + 2\text{BF}_3 \quad (2.1)$$

EIS spectra show the R_{ct} value of 209 Ω and 137 Ω for h-BNNSs and F-h-BNNSs, respectively [50]. From this, it is evident that fluorine doping caused a decrease in the band gap of h-BNNS and showed increased conductivity than pristine h-BNNS.

There is no doubt that the BN has more potential properties to be incorporated in battery cathode materials as highlighted in Table 2.1 on different types of devices.

2.3 BN in Anode Materials for Batteries

2.3.1 BN as Efficient Anode for Lithium-Ion Battery

A lithium-ion battery (LIB), one of the promising energy storage devices, requires considerable research on anode materials to obtain superior stability and cyclability to meet the increasing energy demand. Boron nitride plays a significant role as an additive anode material for various types of batteries. Despite its superior chemical stability, high thermal conductivity, and mechanical properties, BN is a poor electrical conductor. In order to increase the conducting nature of BN, the addition of binders and conductive additives is necessary. However, these additives and coatings on current collectors lead to a decrease in the overall electrical and electrochemical performance of the battery [54]. Therefore, various techniques like electrospinning, facile vacuum filtration, and in situ polymerization have been adopted to synthesize binder-free BN anode materials for LIB [54–56]. The electrospinning technique was used to synthesize carbon nanofiber/boron nitride composites (CNF/BN) with 3, 5, and 7 wt% of BN as flexible anodes for LIB without any binder and current collector [54]. These composites show high electrical conductivity, flexibility, porous structures to accommodate Li^+, reduced Li^+ transport distance, an increased contact area between electrode and electrolyte, and exhibited reversible capacity of 205 mAh g^{-1} after 300 cycles, at 100 mA g^{-1} with a specific amount of BN (3 wt%) content. Li et al. [55] prepared rGO/BN nanosheets composite film with BN contents of 0, 2, 5, and 11 wt% using facile vacuum filtration for binder-free anode for LIB. A combination of rGO and BN leads to enhanced electrolyte access and diffusion to the electrodes and suppresses excessive interlayer expansion due to lithiation and good utilization of active materials. This is mainly because the restacking effect of rGO is inhibited by the BN material in the composite. The rGO/BN composite film shows a high-reversible capacity larger than that of the pure rGO, from which the inhibition of restacking effect by adding BN was seen clearly [55]. This leads to the superior electrochemical properties of rGO/BN-2% composite film, showing a high reversible capacity of 278 mAh g^{-1} at a current density of 100 mA g^{-1} with no obvious decay after 200 cycles. Similarly, Abass et al. [56] prepared polymer-derived silicon oxycarbide SiOC–BNNT/rGO composite with 2, 0.5, and 0.25 wt% BN loading as a binder-free anode electrode. Here the pathway for electron flow and structural support were provided by rGO, and the composite with 0.25 wt% BNNT exhibited good electrochemical performance and it demonstrated the highest capacitance when tested for symmetric supercapacitor application. It is important to note from these reports that the increased wt% of BN leads to poor electrochemical performance, which is a characteristic of the insulating nature of BN. Therefore, appropriate BN loading is needed for the BN-based electrode materials for specific applications. When compared to graphite electrodes, BN has high voltage hysteresis and no flat potentials at lower voltages, which limits bare BN for batteries applications.

Table 2.2 Electrochemical performance of the BN as anode material

Composition	Method	Device	Voltage range	Current density	Capacity mAh g^{-1}	Cycle	Capacity/ Retention (%)	Ref.
CNF/BN-3%	Electrospinning	LIB	0.0–3.0	100 mA g^{-1}	288	300	205/84	[54]
rGO/BN-2%	Vacuum filtration	LIB	0.0–3.0	100 mA g^{-1}	278	200	247/–	[55]
SiOC–BNNT-0.25 wt%	Pyrolysis	LIB	0.0–2.5	100 mA g^{-1}	410	25	238/–	[56]
F-BN	Melt-solid reaction	LIB	0.01–3.0	27 mA g^{-1}	400	–	–	[57]
Carbon-encapsulated hBN@Fe_3O_4	In situ polymerization	LIB	0.0–3.0	200 mA g^{-1}	587	200	512/–	[58]
$Li_4Ti_5O_{12}$/r-GO/hBN	Annealing	LIB	0.5–3.0	20 C	174.1	100	–/85	[59]
Co_3O_4/RGO/hBN	Microwave irradiation	LIB	0.0–3.0	1250 mA g^{-1}	286.04	100	185.3/–	[60]
αP/c-BN/pGra	Freeze-drying	SIB	0.0–1.8	50 mA g^{-1}	1350	100	1000/90	[62]
SiCN-BN/rGO	Pyrolysis	LIB	0.0–2.5	100 mA g^{-1}	~517	1000	~401/–	[63]

Table 2.2 shows the electrochemical performance of the BN as an anode material, which shows the good initial capacity and excellent cyclic performance. Zhang et al. [57] carried out melt-solid reactions of molten LiOH and solid h-BN, which produced agglomerated particles of HO-BN, O-BN, LiO-BN, and H-BN as products of this reaction. When this simultaneously exfoliated and functionalized agglomerated h-BN (FBN) is used as an active material, charge and discharge specific capacity of 78 mAh g^{-1} and 400 mAh g^{-1}, respectively, was obtained by a redox reaction. DFT studies suggest that this redox reaction is caused because of the Li reacting with surface-bound oxygen radicals of O-BN and HO-BN, but not been observed experimentally [57].

Transition metal oxides like Fe_3O_4, $Li_4Ti_5O_{12}$ (LTO), and Co_3O_4 developed along with boron nitride were used to improve the electrochemical performance of LIB and replace the conventional graphite anodes [58, 59]. $Li_4Ti_5O_{12}$ suffers serious disadvantages like poor intrinsic conductivity (10^{-9} S cm^{-1}) and lithium-ion transfer capability. This can be overcome by developing LTO/r-GO/h-BN composites. On comparing the electrochemical performance of the cells with pure LTO, LTO/r-GO, and LTO/r-GO/h-BN composites, LTO/r-GO/h-BN composites exhibited ionic diffusion abilities, better charge transfer, high reversible capacity (at a charging rate of >5C), improvement in anodic stability, and high electrochemical performance. This is because, in r-GO/h-BN modified composites, r-GO leads to electrical conductivity improvement and much better ohmic resistance; whereas, h-BN leads to good charge transfer behavior due to better polarization abilities [59]. In Co_3O_4/ rGO/h-BN transition metal oxide nanocomposites, the addition of h-BN prevented the thermal runaway by enhancing the thermal stability up to 300 °C. Because of this, a 100% capacity retention and Coulombic efficiency were obtained at 150 °C [60]. Interestingly, Duan et al. [58] used h-BN nanosheets as 2D substrates to load carbon-encapsulated 0D Fe_3O_4 nanoparticles, which form a hybrid architecture. Here, the h-BN nanosheets served as a flexible substrate (Fig. 2.4), which accommodates the volume change in Fe_3O_4 by opening the structure of h-BN nanosheets leading to the formation of nanosized channels for electrolyte penetration during repeated cycles. This is because h-BN nanosheets exist in the unsolved state, due to the uncompensated huge surface energy of 2D crystals, which leads to restacking. When Fe_3O_4 is added, spontaneous attraction toward the uncompensated h-BN through van der Waals interactions takes place, which separates them from restacking by acting as spacers [58]. Although Co, Fe-oxides/BN composites have better capacity retention for practical applications, further reductions in discharge/charge potentials are needed [61].

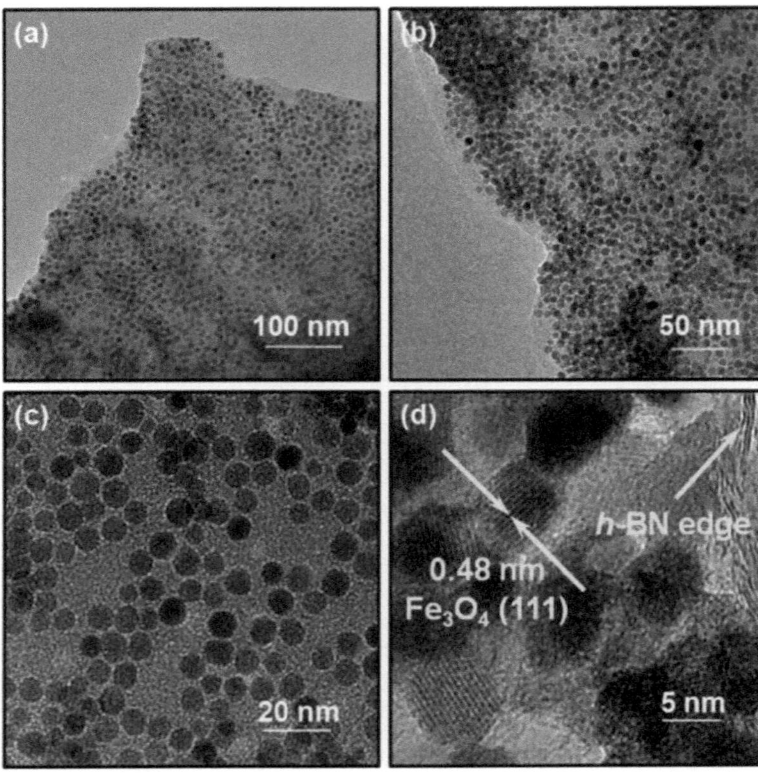

Figure 2.4 (a–c) TEM and (d) HRTEM images of h-BN@Fe_3O_4 hybrid architecture [57].

2.3.2 BN as Efficient Anode for Sodium-Ion Battery

Sodium-ion battery (SIB) due to its low cost, environment-friendly, and abundant sodium (Na) source has gained attention for large-scale energy storage. To enhance the fast transport of Na^+ ions and electrons and also to suppress the volume expansion during sodiation, provided with a shielding layer to offer an excellent elastic buffer backbone a composite with amorphous P (αP), cubic boron nitride (c-BN) wrapped by nanoporous graphene (pGra) networks (αP/c-BN/pGra composite) was developed using ball-milling and freeze-drying methods [62]. In this composite, the nano-miller c-BN, which is an insulator with a rigid skeleton structure, has been coated with black phosphorous (bP) during the ball milling. During this process, the layered structured and highly conducting bP converts the c-BN to an electrically conductive rigid skeleton structure. Hence, due to the synergetic effects of pGra and c-BN modification, the enhanced electrochemical performance of the anode was obtained with a high capacity of around 1000 mAh g^{-1} at 50 mA g^{-1} and 90% capacity retention after 100 cycles [62].

2.4 BN for Electrolyte/Separator in Batteries

Apart from electrodes, the separator also plays a major role in the performance of a battery by preventing contact between the electrodes and short-circuiting [61]. At the same time, the prime role of the separator is to conduct the ion species when it acts as a separator/solid electrolyte. Due to ultrathin structural features, mechanical strength, high thermal stability, and high chemical stability, h-BN has attained great research interest for the electrolyte/separator. Coating commercial separators with boron nitride show greater properties than pure commercial separators [64–66]. Coating density, the number of coating sides (one side or both sides), and wt% of BN-loaded in coating solutions play a significant role. Figure 2.5 shows the fire retardation test of the Celgard-PP separator and BNNT-modified PP separator - one-side coated and double-side coated. The flame retardation test proved the negligible thermal shrinkage of double-side coated BNNT-modified PP separator [64]. BN-modified commercial separators show higher ionic conductivity, which reflects in good Li-ion conduction, due to its good wettability. Although BN exhibits a high hydrophobic nature, the preparation of BN coating solutions with proper binders introduces a hydrophilic nature into the BN separator so that high electrolyte uptake is achieved [64]. The BN-modified commercial separators exhibit superior thermal stability, high thermal conductivity, fire resistance, high capacity, high-rate capability, excellent capacity retention, and negligible self-discharge ($<1\%$), and restrict dissolved polysulfide migration [65, 66] (in the case of Li-S battery).

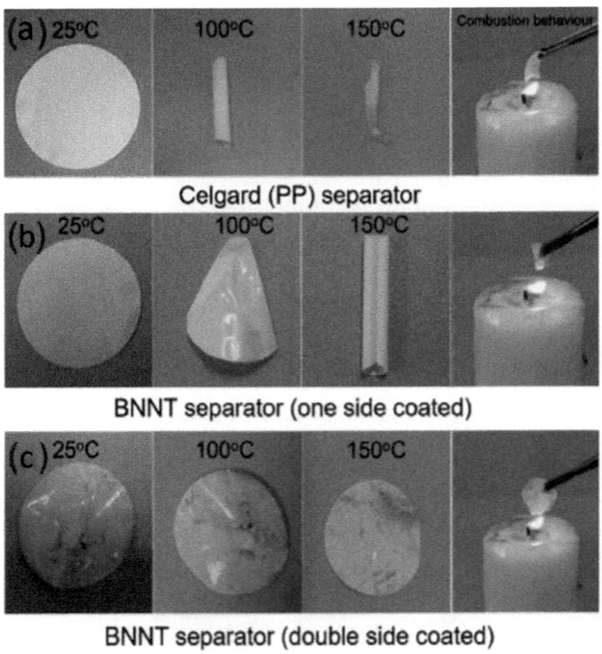

Figure 2.5 (a–c) Evaluation of thermal shrinkage over a temperature ranges from 25 to 150 °C with combustion behavior (most righthand side): (a) Celgard-PP separator, (b) BNNT separator (one-side coated), and (c) BNNT separator (double-side coated) [64].

Polymer/boron nitride composites are reported with various polymers like poly (vinylidene fluoride-co-hexafluoro-propylene) (PVDF-HFP), polyethylene (PE), polyacrylonitrile (PAN), polyethylene oxide (PEO), poly (butyl methacrylate) (PBMA), etc. Various techniques like solution casting, wet chemistry, electrospinning, dry phase-inversion, thermal polymerization, doctor blade, 3D printing, vacuum filtration, dry-pressing, in situ vulcanization, etc. are used to prepare polymer/boron nitride composites. The addition of BN into the PVDF-HFP polymer matrix abruptly disturbed the ordered fashion of polymer chains leading to the suppression in crystallinity and increased amorphous nature.

Table 2.3 Electrochemical performance of BN for electrolyte/separator in batteries

Membrane	Method	Battery type	Voltage range (V)	Current rate	Capacity (mAh/g)	Cycle	Capacity/ retention (%)	Ref.
Celgard PP-BNNT	Doctor blade	LIB	4.3–2.5	1 C	130	–	– / –	[64]
FBN/G	In situ vulcanization	Li-S	2.8–1.6	3 C	1100	1000	700/ –	[65]
Celgard -BN	Filtration	Li-S	1.5–3.0	0.1 C	1210	70	1121	[66]
0.5 wt% BN- (PVDF-HFP)	Solution casting	LIB	2.0–4.5	0.2 C	263	300	– /79	[67]
PEO/LiTFSI/h-BN	Solution casting	LIB	2.8–4.0	0.2 C	143.3	140	134/93	[68]
BN doped $Li_7P_3S_{11}$	Annealing	Li-S	1.0–3.0	C/20	646.5	50	185.4/ –	[69]
BN-PVDF-HFP	3D Printing	LIB	2.5–4.2	0.2 C	139.4	–	–	[70]
PE-BN/PVDF-HFP	Wet chemistry	LIB	2.5–4.2	2 C	120	500	–/95	[71]
h-BNNS/PVDF	Dry phase-inversion	LIB	1.0–2.5	0.1 C	162	100	158/–	[72]
10 wt% hBN/PAN	Electrospinning	LIB	2.5–4.2	0.1 C	144	50	–/92.4	[73]
PVDF-HFP/LiTFSI-1%BN	Doctor-blade	LIB	2.5–4.4	0.2 C	150	50	116/~80	[74]
hBN/EMIM-TFSI/LiTFSI	Glass microfiber filter	LIB	1.0–2.5	0.1 C	165	200	159/96	[75]
BN- 1M LiTFSI/RTIL	Solution casting	LIB	1.0–3.0	C/8	158	40	–/~97	[76]

Membrane	Method	Battery type	Voltage range (V)	Current rate	Capacity (mAh/g)	Cycle	Capacity/ retention (%)	Ref.
AFBNNSs-ion gel	Thermal polymerization	LIB	2.5–4.2	0.2 C	132.8	60	122.4/92.2	[77]
hBN	Doctor blade	LIB	3.6–4.4	0.5 C	~95	100	–/82	[78]
G-CFBN	Doctor blade	LIB	2.5–4.2	10 C	100	500	82/82	[79]
FBN-Celgard 2400	Vacuum filtration	Li-S	1.6–2.8	0.2 C	1485	100	1250/83	[80]
BN-P@GO	Freeze drying	Li-S	1.6–2.8	1 C	1045.3	500	728.7/–	[81]
BNNS/LiTFSI	Dispersion	Li-S	1.5–2.7	0.2 C	~1000	200	881/	[82]
BN-carbon	Doctor blade	Li-S	1.5–3.0	0.5 C	1018.5	250	780.7/–	[83]
PIL-IL-BNNS	Solution casting	Li-metal	2.6–4.2	0.2 C	150	200	114.4/99.4	[84]
BNNSs-coated PEO	Dry-pressing	Li metal	2.5–4.0	2 C	119	200	110/ –	[85]
PVdF-HFP-PBMA-BN	Solution casting	Na-S	0.5–2.5	500 mA/g	656	100	570/88	[86]
g-BN-Li-IL	Squash technique	Polymer battery	2.0–4.0	0.1 C	149	–	–	[87]

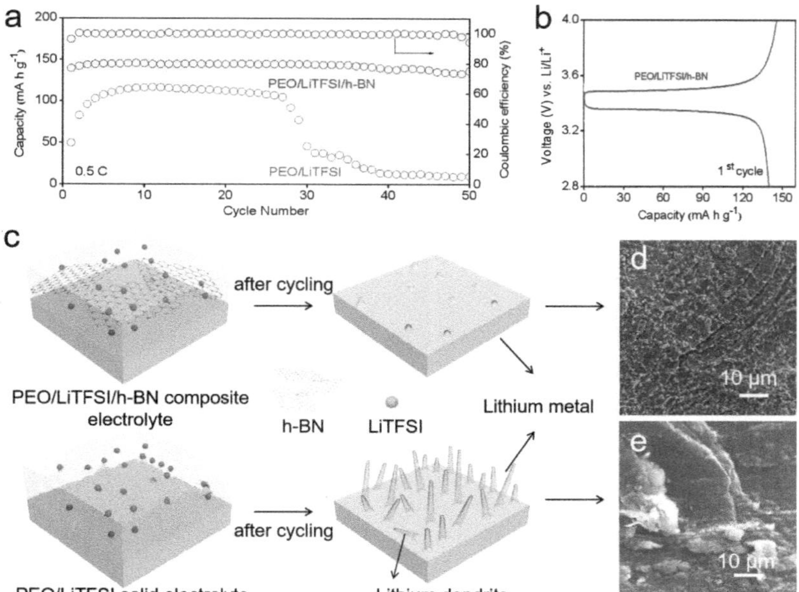

Figure 2.6 (a) Cycle performance and Coulombic efficiency of Li/PEO/LiTFSI/6% h-BN/$LiFePO_4$ and Li/PEO/LiTFSI/$LiFePO_4$ battery at 0.5 C at 60 °C. (b) Charge/discharge voltage curves for 1st cycle of Li/PEO/LiTFSI/6% h-BN/$LiFePO_4$ and the battery using PEO/LiTFSI SPE at 0.5 C. (c) Schematic showing the lithium dendrite inhibition of PEO/LiTFSI SPE PEO/LiTFSI/6% h BN CPE after cycling. (d) The SEM images: Top view of Li anode after 50 cycles at 0.5 C with PEO/LiTFSI/6% h-BN CPE. (e) Top view of Li anode using PEO/LiTFSI SPE [68].

This amorphous nature increases with increasing BN addition, which helps in improved ion movement. Even though the high content of BN increases the ion movement, excessive amounts of BN additive easily cause particle agglomeration and hinder Li-ion transportation [67]. The first principal calculation validated that the large adsorption energy of h-BN suppressed the growth of Li dendrites. Li et al. [68] prepared PEO/LiTFSI/h-BN composite polymer electrolyte by the solution casting method to study the Li dendrite inhibition. Li electrodes after cycling with PEO/LiTFSI/h-BN and PEO/LiTFSI in the battery are examined by SEM. The battery with PEO/LiTFSI/h-BN showed a smooth surface when compared to PEO/LiTFSI, which had a growth of dendrites on it (Fig. 2.6). The battery with PEO/LiTFSI/

h-BN showed a capacity retention of 93% after 140 cycles [68]. Xu et al. [69] prepared BN-doped $Li_7P_3S_{11}$ solid electrolytes to study the interfacial compatibility between electrodes and electrolytes. When $Li_7P_3S_{11}$ is doped with BN, wrapping of BN flakes on the $Li_7P_3S_{11}$ occurs, which protects the interaction of $Li_7P_3S_{11}$ with the Li metal. The increase in doping wt% of BN increases the interfacial compatibility but it suppresses the Li-ion transportation due to agglomeration [69].

2.5 BN for High-Performance Supercapacitors

2.5.1 BN for Improved Efficiency of Asymmetric Supercapacitors (ASC)

Though BN is an insulator, still it can be utilized for enhancing the electrochemical performances of ASC devices. This is possible by tuning the band gap of the BN (4–6 eV) by the formation of heterostructure superlattices. The stacking of different 2D crystals on top of each other leads to the formation of heterostructures and the stacks are bonded with weak van der Waals-like forces [10, 88]. The stacking of these heterostructures leads to the formation of the superlattice. When the heterostructures are made between BN and carbonaceous materials (GO, RGO, graphene, CNT), the difference in the electro-negativity between the boron, nitrogen, and the carbon atom leads to inversion in symmetry and enhances the availability of free electrons [89]. This leads to an increase in charge carrier mobility. Accordingly, when BN and carbonaceous materials are used for electrode materials, the heterostructure superlattice is formed [89]. The band structure and electronic work functions are changed by the redox charge storage mechanism from BN and carbonaceous materials. In this heterostructure superlattice, the role of BN is to reduce the restacking effect of carbonaceous materials, alter the morphology of the surface, and enhance the electrochemical performance [90]. Saha et al. [91–95] reported various studies on rGO/hBN materials for asymmetric supercapacitor applications synthesized by different techniques and studied the band structure modifications, transition from n- to p-type semiconductor by fluorine doping, and the electrochemical properties. The band structure of the materials greatly influences the electronic and electrochemical properties. The band gap of h-BN (4–6 eV) was reduced to ~2 eV upon the addition of rGO, which also shows the formation of a homogeneous crystalline composite [91, 93–94]. It can be observed that the band gap decreases with the increasing content of rGO, which reflects on the notable increase in electrical conductivity. Further decrease in the band gap of rGO/hBN composite materials was achieved by doping of fluorine (F) atom through a soft chemical route [94]. The F-atom replaced B or N in h-BN and C atoms in rGO, which leads to two and three holes, respectively, in the valence band; whereas, the conduction band was left undisturbed. With this, the band gap reduced to 1.77 eV for 7.63 atomic% of F doping. Figure 2.7 shows the XPS valance band spectra, valance band offset, UV-vis spectra and $(\alpha h\nu)^2$ vs. hν plot of the 0, 1, 1.5, and 2 g of F-doped rGO/hBN composite.

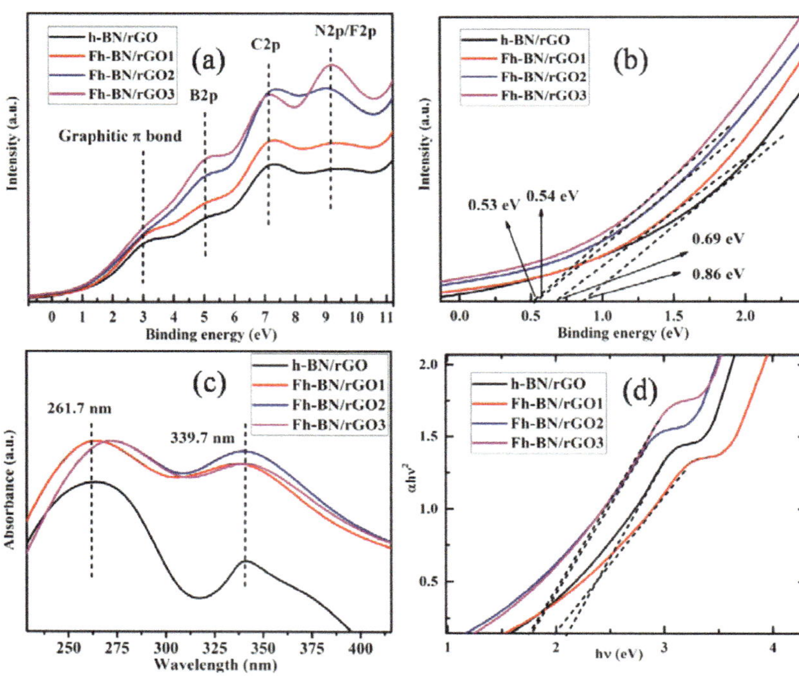

Figure 2.7 (a) XPS valance band spectra, (b) valance band offset, (c) UV-vis spectra, and (d) (αhν) 2 vs. hν plot of h-BN/rGO, Fh-BN/rGO1, Fh-BN/rGO_2, and Fh-BN/rGO_3 [94].

When the asymmetric supercapacitor was fabricated using 1.5 g of F-doped rGO/hBN as positive and thermally reduced graphene oxide (TRGO) as a negative electrode, the specific capacitance was 130% higher than the un-doped samples and exhibited ~77% stability after 10,000 charge-discharge cycles [94]. When other carbonaceous materials like carbon, graphene, and CNT are composited with BN, excellent cyclic stability of ~99%, 96%, and 87%, respectively, are obtained after long cycles. Table 2.4 shows the preparation methods of the BN composites with various carbonaceous materials and the electrochemical performance of the fabricated ASC device.

2.5.2 BN for Improved Efficiency of Symmetric Supercapacitors

There are only a few works reported on BN as an electrode material for symmetric supercapacitor applications (Table 2.5). Among them, most of the reports used a combination of rGO and boron nitride synthesized using various methods. Yang et al. [98] synthesized few-layer h-BNNS/rGO composites by the hydrothermal method and investigated the wide-temperature-range operation with robust stable cycling performance. In neutral aqueous electrolytes (1M Na_2SO_4), capacitance retention of 107, 108, and 111% was observed at temperatures 0 °C, 25 °C, and 50 °C, respectively, after 5000 cycles. The prepared electrode material exhibited robustness to wide-temperature-range operations from −10 to 50 °C, due to the combination of two (h-BNNS and rGO) 2D van der Waals-bonded materials, which caused the cut down of the band gap of the h-BNNSs by rGO, and temperature stability of GO is enhanced by few-layer h-BNNSs. Gao et al. and Byun et al. [99–100] used liquid-phase exfoliation and vacuum filtration techniques, respectively, to synthesize BN/rGO. Gao et al. [99] reported a specific capacitance of 140 F g^{-1} at 2 A g^{-1}, a good rate capability of 71.5 F g^{-1} at 50 A g^{-1}, and the capacitance increases by 5.5% after 1000 cycles, which show excellent cycle stability and low resistance. Byun et al. [100] synthesized BN/rGO hybrid films having heterostructure from OH-BNNP and NH_2-rGO where OH-BNNP serves as mechanical supports and as an electrolyte channel to the rGO.

Table 2.4 Electrochemical performance of BN as an electrode for asymmetric supercapacitors

Electrode	Electrolyte	Method	Current density/ scan rate	Specific capacity ($F\ g^{-1}$)	Power density ($W\ kg^{-1}$)	Energy density ($Wh\ kg^{-1}$)	Stability (%)/ cycle	Ref.
BN/CNT/PANI//rGO	1M KCl	Polymerization	$1\ A\ g^{-1}$	30	400	34.44	87/6000	[89]
m_KBN/CNT/PANI//CB	1M KCl	Sonication	$2\ A\ g^{-1}$	111	–	62	~91/5000	[90]
h-BN/r-GO//TRGO	6M KOH	Hydrothermal	$6\ A\ g^{-1}$	145.7	~4200	39.6	89/4000	[91]
SAA@rGO/hBN@Ni//TRGO	6M KOH	Electrodeposition	$3\ A\ g^{-1}$	268	~7200	~95.3	~80/10000	[92]
h-BN/r-GO//TRGO	6M KOH	Simple annealing	$1\ A\ g^{-1}$	270	14000	73	~80/10000	[93]
Fh-BN/rGO_2//TRGO	6M KOH	Soft chemical route	$3\ A\ g^{-1}$	324	6300	87.5	~77/10000	[94]
Fe_3O_4/NiO@rGO/h-BN//TRGO	KOH	Electro precipitation	$10\ mV\ s^{-1}$	1328	5600	82	79/10000	[95]
h-BN/C//AC	2M KOH	Simple annealing	$0.5\ A\ g^{-1}$	83	245	17	~99/1000	[96]
BN/G//AC	6M KOH	Liquid phase exfoliation	$10\ A\ g^{-1}$	134	1998.5	2.05	96/10000	[97]

Table 2.5 Electrochemical performance of BN as an electrode for symmetric supercapacitors

Electrode	Electrolyte	Method	Current density/ scan rate	Specific capacity ($F g^{-1}$)	Power density	Energy density	Stability (%)/cycle	Ref
SiOC–BNNT/rGO	6M KOH	Pyrolysis	$1 A g^{-1}$	78.93	–	–	86/185	[56]
h-BNNSs/rGO	1M Na_2SO_4	Hydrothermal	$1 A g^{-1}$	42.7	–	–	108/ 5000	[98]
BN/rGO	6M KOH	Liquid-phase exfoliation	$2 A g^{-1}$	140	–	–	105.5/ 1000	[99]
BN/rGO	6M KOH	Vacuum filtration	10 $mV s^{-1}$	242 $F cm^{-3}$	129 $W cm^{-3}$	8.41 $mWh cm^{-3}$	99/10000	[100]
$mSiO_2@Ni_3S_2/NiS_2$	6M KOH	Hydrothermal	$1 A g^{-1}$	370.37	1000 $W kg^{-1}$	41.67 $Wh kg^{-1}$	199/ 2500	[101]

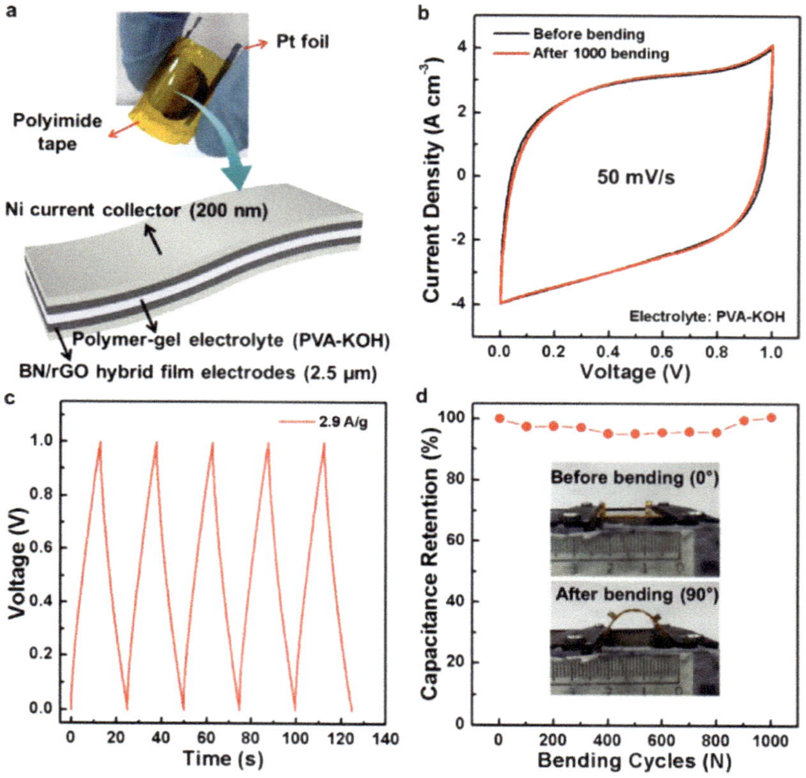

Figure 2.8 BN/rGO hybrid film-based flexible supercapacitor. (a) Schematic diagram of an all-solid-state BN/rGO hybrid film-based flexible supercapacitor and a digital photograph showing the flexibility of the device. (b) Electrochemical properties of the optimized hybrid film (0.5:4 BN/rGO)-based flexible supercapacitor. CV curves of the hybrid film-based device before and after 1000 bending cycles, which show almost the same integrated area after the bending cycles. (c) Charge/discharge curve of the film at a current density of 2.9 A g^{-1} without any significant IR drop. (d) Mechanical properties of the neat NH_2-rGO and hybrid films demonstrate that the optimal amount of BN in the hybrid film acted as a mechanical nanofiller and enhanced the tensile strength of the film [100].

Upon various mass ratios of OH-BNNP:NH_2-rGO varied from 0.25:4 to 2:4, 0.5:4 BN/rGO films retained ~99% of its initial capacitance even after 10000 cycles revealing a superior cyclic stability. When this membrane was tested for flexible supercapacitors, high volumetric capacitance, superior rate capability, and permanent cycle life were achieved. Figure 2.8 shows the performance of the BN/rGO hybrid film-based flexible supercapacitor. The most important factors to evaluate the electrochemical performance of the supercapacitor device are energy density and power density. A porous $mSiO_2@Ni_3S_2/NiS_2$ nickel-based sulfide electrode material was synthesized by Song et al. [101] to develop high-energy-density supercapacitors. The as-synthesized electrode when assembled to an aqueous symmetric supercapacitor showed a specific capacity of 370.37 F g^{-1} at 1 A g^{-1} and ultra-high-energy density of 41.67 Wh kg^{-1} at a power density of 1000 W kg^{-1}. Even after 2500 cycles at 6 A g^{-1}, capacitance retention of 199% was seen, which showed the good rate capability of the material. This high performance was due to the (a) presence of both mesoporous and microporous structures, which helps in the diffusion of ions and sufficient electroactive sites, (b) an increase in the electron transfer efficiency due to the presence of BN and carbon, and (c) $mSiO_2$ that gives the stability for cycling stability.

2.5.3 BN as Polymer Electrolyte Separators for Supercapacitors

Gilshteyn et al. [102] made a comparative study on commercial Celgard separators and the BNNT separators prepared through the airbrush technique. Two-electrode cells were fabricated and at the electrochemical performance were recorded at 25 A g^{-1} in the voltage range of 0–0.8 V. It is noted that the BNNT separator showed high-specific capacitance (82 F g^{-1}) than the commercial (76 F g^{-1}) separator. When a prototype of a stretchable supercapacitor with a BNNT separator is made, the separator withstands at least 1000 cycles with 50% strain and exhibits a specific capacitance of 8.4 F g^{-1}, which is mainly because of the better diffusion of electrolytes during stretching. Zheng et al. [103] fabricated planar asymmetric supercapacitors based on an all-in-one monolithic film to enhance the flexibility, shape diversity, volumetric energy density, and extremely short ionic pathway. This also avoids the use of two metal-based current collectors, polymer binders, conductive additives, and conventional separators, greatly simplifying the device fabrication process. The monolithic film has a sandwich-like layered heterostructure of MnO_2-PH1000/BN nanosheet/exfoliated graphene coated on PET substrate with PVA/LiCl gel electrolyte (EG//MP-PASCs), which exhibits high and superior areal capacitance of ~9.6 mF cm^{-2} and volumetric energy density of 8.6 mWh cm^{-3}. This was relatively high when compared to other conventional asymmetric and planar symmetric supercapacitors. Further, when EG//MP-PASCs were bent at 180°, almost 1000 times, it showed 95.4% capacitance retention. This proved wonderful electrochemical stability and robust mechanical flexibility [103]. Figure 2.9 shows the flexibility of all-solid-state EG//MP-PASCs. For mass production of printable planar asymmetric energy storage devices, this would be a highly recommended technique. Hu et al. [104] made a novel study on ionic conductivity by adding (0.01, 0.025, 0.05, 0.1, 0.25, and 0.5 mg/ml) h-BN nanosheets as fillers to the PVA-H_2SO_4 GPEs. An impressive ionic conductivity of 29 mS/cm was obtained for 0.025 mg/ml h-BN added PVA-H_2SO_4 GPE; whereas, it was only 9 mS/cm for pure PVA-H_2SO_4 GPE. The restacking effect of the h-BN sheets was observed when 0.1 mg/ml h-BN was added, which was evident from the sharp decrease in conductivity (1 mS/cm). Notable electrochemical performance was observed exhibiting capacitance retention of 99.2% after 5000 cycles with nearly 100% columbic efficiency. This shows that adding of precise amount of h-BN leads to low charge transfer resistance, leading to the "super highway" for ion conduction [104]. Gunday et al. [14] also inferred that the addition of a precise amount of h-BN influences the electrical and electrochemical properties of supercapacitors. The addition of h-BN (0.01 g) and ionic liquids (0.2 g) to the polymer showed good ionic conductivity of 1.87×10^{-4} S/cm at 30 °C and 1.3×10^{-3} S/cm at 100 °C, which was due to the strong plasticizing effect of the ionic liquid [14].

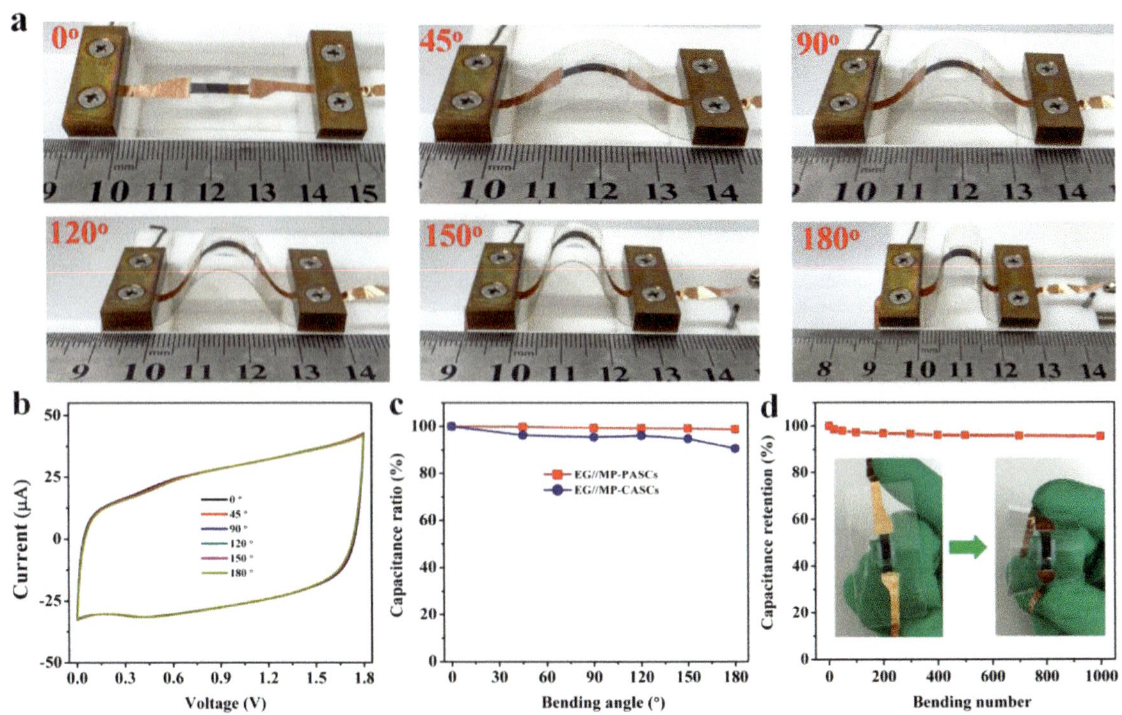

Figure 2.9 Flexibility of all-solid-state EG//MP-PASCs. (a) Photographs of EG//MP-PASCs tested under different bending states of 0–180°. (b) CV curves of EG//MP-PASCs measured at different bending angles at 100 mV s^{-1}. (c) Capacitance ratio as a function of bending angle of EG//MP-PASCs and EG//MP-CASCs. (d) Cycling stability of EG//MP-PASCs tested under repeated bending state. Inset is the photograph of EG//MP-PASCs under flat and bending states [103].

Table 2.6 Electrochemical performance of BN as polymer electrolyte separators for supercapacitors

Material	Method	Voltage range (V)	Energy density	Power density	Current density/ scan rate	Capacity (F g^{-1})	Retention (%) /Cycle	Ref.
SPSU/%5hBN/0.2 IL	Solution casting	0.1–1	43.8 Wh kg^{-1}	1100 W kg^{-1}	1 A g^{-1}	90.4	99/5000	[14]
BNNT	Airbrush techniques	0–0.8	3.7 mWh cm^{-3}	6200 m W cm^{-3}	25 A g^{-1}	82	96/20000	[102]
EG//MP-PASCs	Ball milling	0–1.8	8.6 m Wh cm^{-3}	4.2 W cm^{-3}	10 mV s^{-1}	19.2	92/5000	[103]
BN-PVA-H_2SO_4	Freeze-thaw method	0–1	–	–	0.5 A g^{-1}	124.5	99.2/5000	[104]

2.6 Conclusion and Future Perspective

Among various 2D materials, boron nitride has gained an attractive research interest, due to its unique properties for electrochemical energy storage devices. Boron nitride is an insulator with a large band gap. Formation of heterostructures and superlattice with other 2D materials and by doping of other materials with BN significantly reduced the band gap of BN. This is significantly reflected in the increase in electronic conductivity. However, the increase in wt% of BN loading leads to agglomeration of particles, which causes a block in the ion transportation and affects the electrochemical performance. Therefore, proper wt% BN should be taken for better electrochemical performance. The addition of BN inhibits the restacking effect when composited with graphene-based materials due to the synergistic effect. BN shows stronger adsorption of polysulfides than graphene and it is an excellent sulfur fixer in sulfur-based batteries. The coating of BN on commercial separators showed superior electrochemical properties to commercial ones. BN also serves as a flexible substrate, which leads to high electrolyte uptake. Therefore, considering the good chemical and thermal stability, BN is a promising 2D material for electrochemical energy storage devices. From this, it can be concluded that boron nitride is a suitable material for commercialization in energy storage applications.

Acknowledgments

The authors Dr. A. Sakutnhala and Dr. S. Rajesh thank the Department of Science and Technology, Science and Engineering Research Board (DST-SERB), (EMR/2017/003227 dated 16 July, 2018), Government of India, for the funding. The authors also thank Karunya Institute of Technology and Sciences, Coimbatore, 641 114, India, for its research facilities.

References

1. M. Li, Z. Luo, Y. Zhao, Recent advancements in 2D nanomaterials for cancer therapy, *Sci. China Chem.*, 61 (2018), 1214–1226. https://doi.org/10.1007/s11426-018-9294-9.

2. M. Z. Iqbal, N. A. Qureshi, G. Hussain, Recent advancements in 2D-materials interface based magnetic junctions for spintronics, *Journal of Magnetism and Magnetic Materials,* 457 (2018), 110–125. https://doi.org/10.1016/j.jmmm.2018.02.084.
3. Y. Li, Y.-L. Li, B. Sa, R. Ahuja, Review of two-dimensional materials for photocatalytic water splitting from a theoretical perspective, *Catal. Sci. Technol.*, 7 (2017), 545–559. https://doi.org/10.1039/C6CY02178F.
4. L. Peng, Y. Zhu, D. Chen, R. S. Ruoff, G. Yu, Two-dimensional materials for beyond-lithium-ion batteries, *Adv. Energy Mater.*, 6 (2016), 1600025. https://doi.org/10.1002/aenm.201600025.
5. W.H. Balmain, XLVI. Observations on the formation of compounds of boron and silicon with nitrogen and certain metals, *The London, Edinburgh, and Dublin Philosophical Magazine and Journal of Science,* 21–138 (1842), 270–277. https://doi.org/10.1080/14786444208621545.
6. J. Yin, J. Li, Y. Hang, J. Yu, G. Tai, X. Li, Z. Zhang, W. Guo, Boron nitride nanostructures: Fabrication, functionalization and applications, *Small,* 12 (2016), 2942–2968. https://doi.org/10.1002/smll.201600053.
7. R. Pease, Crystal structure of boron nitride, *Nature,* 165 (1950), 722–723. https://doi.org/10.1038/165722b0.
8. X. Jiang, Q. Weng, X. Wang X. Li, J. Zhang, D. Golberg, Y. Bando, Recent progress on fabrications and applications of boron nitride nanomaterials: A review, *Journal of Materials Science & Technology,* 31 (2015), 589–598. https://doi.org/10.1016/j.jmst.2014.12.008.
9. J. Wang, F. Ma, W. Liang, M. Sun, Electrical properties and applications of graphene, hexagonal boron nitride (h-BN), and graphene/h-BN heterostructures, *Materials Today Physics,* 2 (2017), 6–34. https://doi.org/10.1016/j.mtphys.2017.07.001.
10. N. Izyumskaya, D.O. Demchenko, S. Das, Ü. Özgür, V. Avrutin, H. Morkoç, Recent development of boron nitride towards electronic applications, *Adv. Electron. Mater.*, (2017), 1600485. https://doi.org/10.1002/aelm.201600485.
11. H. X. Jiangz, J. Y. Lin, *ECS Journal of Solid State Science and Technology,* 6 (2017), Q3012–Q3021. https://doi.org/10.1149/2.0031702jss.
12. B. Ertug, Powder preparation, properties and industrial applications of hexagonal boron nitride. In *Sintering Applications,* Tech: Rijeka, 2013, Chapter 2, 33–55.
13. Y. Yi, H. Li, H. Chang, P. Yang, X. Tian, P. Liu, L. Qu, M. Li, B. Yang, H. Li, W. Zhu, S. Dai, Few-layer boron nitride with engineered nitrogen

vacancies for promoting conversion of polysulfide as a cathode matrix for lithium-sulfur batteries, *Chem. Eur. J.* https://doi.org/10.1002/chem.201900884.

14. S. T. Gunday, E. Cevik, A. Yusuf, A. Bozkurt, Nanocomposites composed of sulfonated polysulfone/hexagonal boron nitride/ionic liquid for supercapacitor applications, *Journal of Energy Storage*, 21 (2019), 672–679. https://doi.org/10.1016/j.est.2019.01.008.
15. L. Lin, T. Liu, Y. Zhang, R. Sun, W. Zeng, Z. Wang, Synthesis of boron nitride nanosheets with a few atomic layers and their gas-sensing performance, *Ceram. Int.*, 42 (2016), 971–975. https://doi.org/10.1016/j.ceramint.2015.08.109.
16. G. Liu, S. L. Rumyantsev, C. Jiang, M. S. Shur, A. A. Balandin, Selective gas sensing with h-BN capped MoS_2 heterostructure thin-film transistors, *IEEE Electron Device Letters*, 36 (2015), 1202–1204. https://doi.org/10.1109/LED.2015.2481388.
17. S. Wan, Y. Yu, J. Pu, Z. Lu, Facile fabrication of boron nitride nanosheets–amorphous carbon hybrid film for optoelectronic applications, *RSC Adv.*, 5 (2015), 19236–19240. https://doi.org/10.1039/C4RA13268H.
18. A. Maity, S. J. Grenadier, J. Li, J. Y. Lin, H. X. Jiang, High efficiency hexagonal boron nitride neutron detectors with 1 cm^2 detection areas, *Appl. Phys. Lett.*, 116 (2020), 142102. https://doi.org/10.1063/1.5143808.
19. S. Mahalingam, S. Ayyaru, Y. Ahn, Enhanced cathode performance of Fe_2O_3, boron nitride-doped rGO nanosheets for microbial fuel cell applications, *Sustainable Energy Fuels*, 4 (2020), 1454–1468. https://doi.org/10.1039/C9SE01243E.
20. N.P. Khalili, R. Moradi, P. Kavehpour, F. Islamzada, Boron nitride nanotube clusters and their hybrid nanofibers with polycaprolacton: Thermo-pH sensitive drug delivery functional materials, *European Polymer Journal*, 127 (2020), 109585. https://doi.org/10.1016/j.eurpolymj.2020.109585.
21. H. Belaid, S. Nagarajan, C. Barou, V. Huon, J. Bares, S. Balme, P. Miele, D. Cornu, V. Cavaillès, C. Teyssier, M. Bechelany, Boron nitride-based nano biocomposites: Design by 3D printing for bone tissue engineering, *ACS Applied Bio Materials*, 3(4) 2020, 1865–1874. https://doi.org/10.1021/acsabm.9b00965.
22. A. Pakdel, Y. Bando, D. Golberg, Nano boron nitride flatland, *Chem. Soc. Rev.*, 43 (2014), 934–959. https://doi.org/10.1039/C3CS60260E.

23. L. Stagi, J. Ren, P. Innocenzi, From 2-D to 0-D boron nitride materials: The next challenge, *Materials*, 12 (2019), 3905. https://doi.org/10.3390/ma12233905.

24. Y. Li, B. Liu, Z. Qin, L. Hou, Influence of milling time on productivity of boron nitride nanotubes synthesized from B_2O_3 by ball milling and annealing method, *2020 IOP Conf. Ser.: Earth Environ. Sci.*, 453 (2020), 012091. https://doi.org/10.1088/1755-1315/453/1/012091.

25. H. Yurdakul, Y. Göncü, O. Durukan, A. Akay, A. T. Seyhan, N. Ay, S. Turan, Nanoscopic characterization of two-dimensional (2D) boron nitride nanosheets (BNNSs) produced by microfluidization, *Ceramics International*, 38 (2012), 2187–2193. https://doi.org/10.1016/j.ceramint.2011.10.064.

26. K. S. Novoselov, A. K. Geim, S. V. Morozov, D. Jiang, Y. Zhang, S. V. Dubonos, I. V. Grigorieva, A. A. Firsov, Electric field effect in atomically thin carbon films, *Science*, 306 (2004), 666–669. DOI: 10.1126/science.1102896.

27. Li Cao, Sadra Emami, Khalid Lafdi, Large-scale exfoliation of hexagonal boron nitride nanosheets in liquid phase, *Mater. Express*, 4 (2014), 165–171. https://doi.org/10.1166/mex.2014.1155.

28. C. F. Kuo, Jeffrey Chen, J. Bo, P. Chen, Dong Yen, M. Yan, Preparation of boron nitride nanosheets using a chemical exfoliation method as a thermal conductive filler for the development of silicone thermal composites, Part I: effect of single- and hybrid-filler additions on the silicone composite performance, *Textile Research Journal*, 90 (2020), 666–684. http://dx.doi.org/10.1177/0040517519877467.

29. Y. Lin, T.V. Williams, T. B. Xu, W. Cao, H. E. E. Ali, J. W. Connell, Aqueous dispersions of few-layered and monolayered hexagonal boron nitride nanosheets from sonication-assisted hydrolysis: Critical role of water, *J. Phys. Chem. C*, 115(6) (2011), 2679–2685. https://doi.org/10.1021/jp110985w.

30. Y. Wang, Z. Shi, J. Yin, Boron nitride nanosheets: Large-scale exfoliation in methane sulfonic acid and their composites with polybenzimidazole, *J. Mater. Chem.*, 21 (2011), 11371–11377. https://doi.org/10.1039/C1JM10342C.

31. Y. Shi, C. Hamsen, X. Jia, K. K. Kim, A. Reina, M. Hofmann, A. L. Hsu, K. Zhang, H. Li, Z. Y. Juang, M. S. Dresselhaus, L. J. Li, J. Kong, Synthesis of few-layer hexagonal boron nitride thin film by chemical vapor deposition, *Nano Lett.*, 10(10) (2010), 4134–4139. https://doi.org/10.1021/nl1023707.

32. P. Ahmad, M. U. Khandaker, Y. M. Amin, Synthesis of boron nitride nanotubes by argon supported thermal chemical vapor deposition, *Physica E*, 67 (2015), 33–37. https://doi.org/10.1016/j.physe.2014.11.003.

33. G. H. Han, J. A. R. Manzo, C. W. Lee, N. J. Kybert, M. B. Lerner, Z. J. Qi, E. N. Dattoli, A. M. Rappe, M. Drndic, A. T. C. Johnson, Continuous growth of hexagonal graphene and boron nitride in-plane heterostructures by atmospheric pressure chemical vapor deposition, *ACS Nano*, 7(11) (2013), 10129–10138. https://doi.org/10.1021/nn404331f.

34. N. R. Glavin, M. L. Jespersen, M. H. Check, J. Hu, A. M. Hilton, T. S. Fisher, A. A. Voevodin, Synthesis of few-layer, large area hexagonal-boron nitride by pulsed laser deposition, *Thin Solid Films*, 572 (2014), 245–250. https://doi.org/10.1016/j.tsf.2014.07.059.

35. Z. F. Zhoua, I. Bello, M. K. Lei, K. Y. Li, C. S. Lee, S. T. Lee, Synthesis and characterization of boron carbon nitride films by radio frequency magnetron sputtering, *Surface and Coatings Technology*, 128–129 (2000), 334–340. https://doi.org/10.1016/S0257-8972(00)00600-9.

36. S. Nakhaie, J. M. Wofford, T. Schumann, U. Jahn, M. Ramsteiner, M. Hanke, J. M. J. Lopes, H. Riechert, Synthesis of atomically thin hexagonal boron nitride films on nickel foils by molecular beam epitaxy, *Appl. Phys.*, 106 (2015), 213108. https://doi.org/10.1063/1.4921921.

37. Z. Gao, C. Zhi, Y. Bando, D. Golberg, T. Serizawa, Noncovalent functionalization of boron nitride nanotubes in aqueous media opens application roads in nanobiomedicine, *Nanobiomedicine*, 1(7) (2014). https://doi.org/10.5772%2F60000.

38. S-Y. Xie, W. Wang, K. A. S. Fernando, X. Wang, Y. Lin, Y-P. Sun, Solubilization of boron nitride nanotubes, *Chem. Commun.*, (2005), 3670–3672. https://doi.org/10.1039/B505330G.

39. H. Shin, J. Guan, M. Z. Zgierski, K. S. Kim, C. T. Kingston, B. Simard, Covalent functionalization of boron nitride nanotubes via reduction chemistry, ACS Nano, 9(12) (2015), 12573–12582. https://doi.org/10.1021/acsnano.5b06523.

40. M. Emanet, Ö. Sen, I. Ç. Taskin, M. Çulha, Synthesis, functionalization, and bioapplications of two-dimensional boron nitride nanomaterials, *Front. Bioeng. Biotechnol.*, 7(363) (2019). https://doi.org/10.3389/fbioe.2019.00363.

41. M. Petrescu, M.-G. Balint, Structure and properties modifications in boron nitride, Part I: Direct polymorphic transformations mechanisms, *UPB Sci. Bull., Series B*, 69 (2007), 35–42.

42. G. R. Bhimanapati, N. R. Glavin, J. A. Robinson, 2D boron nitride: Synthesis and applications, Ch 3, *Semiconductors and Semimetals*, 95 (2016), 101–147, Edited by F. Iacopi, J. J. Boeckl, C. Jagadish.

43. M. V. Reddy, A. Mauger, C. M Julien, A. Paolella, K. Zaghib, Brief history of early lithium-battery development, *Materials*, 13 (2020), 1884. https://doi.org/10.3390/ma13081884.

44. G. Cohn, L. Ma, L. A. Archer, A novel non-aqueous aluminum sulfur battery, *J. Power Sources*, 283 (2015), 416–422. https://doi.org/10.1016/j.jpowsour.2015.02.131.

45. T. Gao, X. Li, X. Wang, J. Hu, F. Han, X. Fan, L. Suo, A. J. Pearse, S. B. Lee, G. W. Rubloff, K. J. Gaskell, M. Noked, C. Wang, A Rechargeable Al/S battery with an ionic-liquid electrolyte, *Angew. Chem. Int. Ed.*, 55 (2016), 1–5. https://doi.org/10.1002/anie.201603531.

46. H. Yang, L. Yin, J. Liang, Z. Sun, Y. Wang, H. Li, K. He, L. Ma, Z. Peng, S. Qiu, C. Sun, H-M. Cheng, F. Li, A fast kinetic response aluminum-sulfur battery, *Angew. Chem. Int. Edit.*, 57 (2018), 1898–1902. https://doi.org/10.1002/anie.201711328.

47. X. Yu, M. J. Boyer, G. S. Hwang, A. Manthiram, Room-temperature aluminum-sulfur batteries with a lithium-ion-mediated ionic liquid electrolyte, *Chem*, 4 (2018), 586–598. https://doi.org/10.1016/j.chempr.2017.12.029.

48. K. Zhang, T. H. Lee, J. H. Cha, R. S. Varma, J-W. Choi, H. W. Jang, M. Shokouhimehr, Two dimensional boron nitride as a sulfur fixer for high performance rechargeable aluminum sulfur batteries, *Sci Rep.*, 9 (2019), 13573. https://doi.org/10.1038/s41598-019-50080-9.

49. D. R. Deng, F. Xue, C-D. Bai, J. Lei, R. Yuan, M. S. Zheng, Quan F. Dong, Enhanced adsorptions to polysulfides on graphene-supported bn nanosheets with excellent Li–S battery performance in a wide temperature range, *ACS Nano*, 12(11) 2018, 11120–11129. https://doi.org/10.1021/acsnano.8b05534.

50. G. Venkateswarlua, D. Madhub, J. V. Rani, An effective performance of F-doped hexagonal boron nitride nanosheets as cathode material in magnesium battery, *Materials Chemistry and Physics*, 226 (2019), 356–361. https://doi.org/10.1016/j.matchemphys.2019.01.051.

51. G. Venkateswarlu, D. Madhu, J. V. Rani, Graphene supported boron nitride nanosheets as advanced electroanalytical performance for rechargeable magnesium storage system, *Chemistry Select*, 5 (2020), 2247–2254. https://doi.org/10.1002/slct.201904872.

52. M. Minakshia, M.G. Blackford, Electrochemical characteristics of B_4C or BN added MnO_2 cathode material for alkaline batteries,

Materials Chemistry and Physics, 123 (2010), 700–705. https://doi.org/10.1016/j.matchemphys.2010.05.041.

53. Z. W. Lu, S. M. Yao, G. R. Li, T. Y. Yan, X. P. Gao, Microstructure and electrochemical properties of the Co–BN composites, *Electrochimica Acta*, 53 (2008), 2369–2375. https://doi.org/10.1016/j.electacta.2007.10.006.
54. M. Chen, S. Zhang, J. Zhang, Q. Chen, Electrospun carbon nanofiber/boron nitride composites as flexible anodes for lithium-ion batteries, *J. Nanosci. Nanotechnol.*, 19 (2019) 220–225. https://doi.org/10.1166/jnn.2019.16446.
55. H. Li, R. Y. Tay, S.H. Tsang, W. Liu, E.H.T. Teo, Reduced graphene oxide/boron nitride composite film as a novel binder-free anode for lithium ion batteries with enhanced performances, *Electrochimica Acta*, 166 (2015), 197–205. https://doi.org/10.1016/j.electacta.2015.03.109.
56. M. A. Abass, A. A. Syed, C. Gervais, G. Singh, Synthesis and electrochemical performance of a polymer-derived silicon oxycarbide/boron nitride nanotube composite, *RSC Adv.*, 7 (2017), 21576–21584. https://doi.org/10.1039/C7RA01545C.
57. F. Zhang, K. Németh, J. Bareño, F. Dogan, I. D. Bloom, L. L. Shaw, Experimental and theoretical investigations of functionalized boron nitride as electrode materials for li-ion batteries, *RSC Adv.*, 6 (2016), 27901–27914. https://doi.org/10.1039/C6RA03141B.
58. Z-Q. Duan, Y-T. Liu, X-M. Xie, X-Y. Ye, X-D. Zhu, h-BN nanosheets as 2D substrates to load 0D Fe_3O_4 nanoparticles: A hybrid anode material for lithium-ion batteries, *Chem. Asian J.*, 11 (2016), 828 – 833. https://doi.org/10.1002/asia.201501439.
59. OnurErgen, Hexagonal boron nitride incorporation to achieve high performance $Li_4Ti_5O_{12}$ electrodes, *AIP Advances*, 10 (2020), 045040. https://doi.org/10.1063/5.0004376.
60. Y. Mussa, F. Ahmed, M. Arsalan, E. Alsharaeh, Two dimensional (2D) reduced graphene oxide (RGO)/hexagonal boron nitride (h-BN) based nanocomposites as anodes for high temperature rechargeable lithium ion batteries, *Sci Rep.*, 10 (2020), 1882. https://doi.org/10.1038/s41598-020-58439-z.
61. M. V. Reddy, G. V. Subba Rao, B. V. R. Chowdari, Metal oxides and oxysalts as anode materials for li-ion batteries, *Chem. Rev.*, 113(7) (2013), 5364–5457. https://doi.org/10.1021/cr3001884.
62. X. Ding, Y. Huang, G. Li, Y. Tang, X. Li, Y. Huang, Phosphorus nanoparticles combined with cubic boron nitride and graphene as

stable sodium-ion battery anodes, *Electrochimica Acta*, 235 (2017), 150–157. http://dx.doi.org/10.1016/j.electacta.2017.03.055.

63. L. David, S. Bernard, C. Gervais, P. Miele, G. Singh, Facile synthesis and high rate capability of silicon carbonitride/boron nitride composite with a sheet-like morphology, *J. Phys. Chem. C*, 119(5) (2015), 2783–2791. https://doi.org/10.1021/jp508075x.
64. M. M. Rahman, S. Mateti, Q. Cai, I. Sultana, Y. Fan, X. Wang, C. Hou, Y. Chen, High temperature and high rate lithium-ion batteries with boron nitride nanotubes coated polypropylene separators, *Energy Storage Materials*, 19 (2019), 352–359. https://doi.org/10.1016/j.ensm.2019.03.027.
65. Y. Fan, D. Liu, M. M. Rahman, T. Tao, W. Lei, S. Mateti, B. Yu, J. Wang, C. Yang, Y. Chen, Repelling polysulfide ions by boron nitride nanosheet coated separators in lithium–sulfur batteries, *ACS Appl. Energy Mater.*, 2 (2019), 2620–2628. https://doi.org/10.1021/acsaem.8b02205.
66. G. Babu, A. Sawas, N. K. Thangavel, L. M. R. Arava, Two-dimensional material-reinforced separator for li–sulfur battery, *J. Phys. Chem. C*, 122 (2018), 10765–10772. https://doi.org/10.1021/acs.jpcc.8b02633.
67. X. Bian, J. Liang, X. Tang, R. Li, L. Kang, A. Su, X. Su, Y. Wei, A boron nitride-polyvinylidene fluoride-co-hexafluoropropylene composite gel polymer electrolyte for lithium metal batteries, *Journal of Alloys and Compounds*, 803 (2019), 1075e1081. https://doi.org/10.1016/j.jallcom.2019.06.321.
68. Y. Li, L. Zhang, Z. Sun, G. Gao, S. Lu, M. Zhu, Y. Zhang, Z. Jia, C. Xiao, H. Bu, K. Xi, S. Ding, Hexagonal boron nitride inducing anionic trapping in polyethylene oxide based solid polymer electrolyte for lithium dendrite inhibition, *J. Mater. Chem. A*, 8 (2020), 9579–9589. https://doi.org/10.1039/D0TA03677C.
69. X. Xu, L. Wang, H. Fei, L. Ci, Boron nitride doped $Li_7P_3S_{11}$ solid electrolyte with improved interfacial compatibility and application in all-solid-state Li/S battery, *J Mater Sci: Mater Electron*, 30 (2019), 19119–19125. https://doi.org/10.1007/s10854-019-02267-z.
70. Y. Liu, Y. Qiao, Y. Zhang, Z. Yang, T. Gao, D. Kirsch, B. Liu, J. Song, B. Yang, L. Hu, 3D printed separator for the thermal management of high-performance Li metal anodes, *Energy Storage Materials*, 12 (2018), 197–203. https://doi.org/10.1016/j.ensm.2017.12.019.
71. M. Waqas, S. Ali, W. Lv, D. Chen, B. Boateng, W. He, High-performance PE-BN/PVDF-HFP bilayer separator for lithium-ion batteries, *Adv.*

Mater. Interfaces, 6 (2019), 1801330. https://doi.org/10.1002/admi.201801330.

72. A. C. M. de Moraes, W. J. Hyun, N.S. Luu, J-M. Lim, K-Y. Park, M. C. Hersam, Phase-inversion polymer composite separators based on hexagonal boron nitride nanosheets for high-temperature lithium-ion batteries, *ACS Appl. Mater. Interfaces*, 12(7) (2020), 8107–8114. https://doi.org/10.1021/acsami.9b18134.
73. H. Aydın, S. Ü. Çelik, A. Bozkurt, Electrolyte loaded hexagonal boron nitride/polyacrylonitrilenano fibers for lithium ion battery application, *Solid State Ionics*, 309 (2017), 71–76. http://dx.doi.org/10.1016/j.ssi.2017.07.004.
74. Z. Zhang, R. G. Antonio, K. L. Choy, Boron nitride enhanced polymer/salt hybrid electrolytes for all-solid-state lithium ion batteries, *Journal of Power Sources*, 435 (2019), 226736. https://doi.org/10.1016/j.jpowsour.2019.226736.
75. W. J. Hyun, A. C. M. de Moraes, J-M. Lim, J. R. Downing, K-Y. Park, M. T. Z. Tan, M. C. Hersam, High-modulus hexagonal boron nitride nanoplatelet gel electrolytes for solid-state rechargeable lithium-ion batteries, *ACS Nano*, 13 (2019), 9664–9672. https://doi.org/10.1021/acsnano.9b04989.
76. M-T. F. Rodrigues, K. Kalaga, H. Gullapalli, G. Babu, A. L. M. Reddy, P. M. Ajayan, Hexagonal boron nitride-based electrolyte composite for li-ion battery operation from room temperature to 150 °C, *Adv. Energy Mater*, 6 (2016), 1600218. https://doi.org/10.1002/aenm.201600218.
77. D. Kim, X. Liu, B. Yu, S. Mateti, L. A. O'Dell, Q. Rong, Y. (Ian) Chen, Amine-functionalized boron nitride nanosheets: A new functional additive for robust, flexible ion gel electrolyte with high lithium-ion transference number, *Adv. Funct. Mater.*, (2020), 1910813. https://doi.org/10.1002/adfm.201910813.
78. A. C. M. de Moraes, W. J. Hyun, J-W. T. Seo, J. R. Downing, J-M. Lim, M. C. Hersam, Ion-conductive, viscosity-tunable hexagonal boron nitride nanosheet inks, *Adv. Funct. Mater*, (2019), 1902245. https://doi.org/10.1002/adfm.201902245.
79. J. Shim, H. J. Kim, B.G. Kim, Y. S. Kim, D-G. Kim, J-C. Lee, 2D boron nitride nanoflakes as a multifunctional additive in gel polymer electrolytes for safe, long cycle life and high rate lithium metal batteries, *Energy Environ. Sci.*, 10 (2017), 1911–1916. https://doi.org/10.1039/C7EE01095H.

80. Y. Fan, Z. Yang, W. Hua, D. Liu, T. Tao, M. M. Rahman, W. Lei, S. Huang, Y. Chen, Functionalized boron nitride nanosheets/graphene interlayer for fast and long-life lithium–sulfur batteries, *Adv. Energy Mater.*, (2017), 1602380. https://doi.org/10.1002/aenm.201602380.

81. J. Zhang, W. Ma, Z. Feng, F. Wu, D. Wei, B. Xi, S. Xiong, P-doped BN nanosheets decorated graphene as the functional interlayer for Li–S batteries, *Journal of Energy Chemistry*, 39 (2019), 54–60. https://doi.org/10.1016/j.jechem.2019.01.016.

82. J. Wu, X. Li, Z. Rao, X. Xu, Z. Cheng, Y. Liao, L. Yuan, X. Xie, Z. Li, Y. Huang, Electrolyte with boron nitride nanosheets as leveling agent towards dendrite-free lithium metal anodes, *Nano Energy*, 72 (2020), 104725. https://doi.org/10.1016/j.nanoen.2020.104725.

83. P. J. H. Kim, J. Seo, K. Fu, J. Choi, Z. Liu, J. Kwon, Li. Hu, U. Paik, Synergistic protective effect of a BN-carbon separator for highly stable lithium sulfur batteries, *NPG Asia Mater*, 9 (2017), e375. https://doi.org/10.1038/am.2017.51.

84. Q. Xia, J. Wu, Q. X. Shi, X. Xiang, X. Li, H. Pei, H. Zeng, X. Xie, Y. S. Ye, UV-curable boron nitride nanosheet/ionic liquid-based crosslinked composite polymer electrolyte in lithium metal batteries, *Journal of Power Sources*, 414 (2019), 283–292. https://doi.org/10.1016/j.jpowsour.2019.01.004.

85. B. Shen, T-W. Zhang, Y-C. Yin, Z-X. Zhu, L-L. Lu, C. Ma, F. Zhou, H-B. Yao, Chemically exfoliated boron nitride nanosheets form robust interfacial layers for stable solid-state Li metal batteries, *Chem. Commun.*, 55 (2019), 7703–7706. https://doi.org/10.1039/C9CC02124H.

86. P. V. K. Saroja, A. Rajamani, K. Muthusamy, R. Sundara, Repelling polysulfides using white graphite introduced polymer membrane as a shielding layer in ambient temperature sodium sulfur battery, *Adv. Mater. Interfaces*, (2019), 1901497. https://doi.org/10.1002/admi.201901497.

87. M. Li, W. Zhu, P. Zhang, Y. Chao, Q. He, B. Yang, H. Li, A. Borisevich, S. Dai, Graphene-analogues boron nitride nanosheets confining ionic liquids: A high-performance quasi liquid solid electrolyte, *Small*, Weinheiman der Bergstrasse, Germany, 26 (2016), 3535–3542. https://doi.org/10.1002/smll.201600358.

88. A. K. Geim, I. V. Grigorieva, Van der Waals heterostructures, *Nature*, 499 (2013), 419–425. https://doi.org/10.1038/nature12385.

89. C. K. Maity, G. Hatui, S. Sahoo, P. Saren, G. C. Nayak, Boron nitride based ternary nanocomposites with different carbonaceous

materials decorated by polyaniline for supercapacitor application, *Chemistry Select*, 4 (2019), 3672–3680. https://doi.org/10.1002/slct.201803560.

90. C. K. Maitya, S. Sahooa, K. Vermab, A. K. Beherac, G. C. Nayak, Facile functionalization of boron nitride (BN) for the development of high-performance asymmetric Supercapacitors, *New J. Chem.*, 44 (2020), 8106–8119. https://doi.org/10.1039/C9NJ06284J.
91. S. Saha, M. Jana, P. Khanra, P. Samanta, H. Y. Koo, N. C. Murmu, T. Kuila, Band gap engineering of boron nitride by graphene and its application as positive electrode material in asymmetric supercapacitor device, *ACS Appl. Mater. Interfaces*, 7(26) 2015, 14211–14222. https://doi.org/10.1021/acsami.5b03562.
92. S. Saha, P. Samanta, N. C. Murmu, N. H. Kim, T. Kuila, J. H. Lee, Electrochemical functionalization and in-situ deposition of the SAA@rGO/h-BN@Ni electrode for supercapacitor applications, *J. Ind. Eng. Chem.*, 52 (2017), 321–330. https://doi.org/10.1016/j.jiec.2017.04.005.
93. S. Saha, M. Jana, P. Samanta, N. C. Murmu, N. H. Kim, T. Kuila, J. H. Lee, Investigation of band structure and electrochemical properties of h-BN/rGO composites for asymmetric supercapacitor applications, *Materials Chemistry and Physics*, (2017). https://doi.org/10.1016/j.matchemphys.2017.01.025.
94. S. Saha, P. Samanta, N. C. Murmu, A. Banerjee, R. Sankar Ganesh, H. Inokawa, T. Kuila, Modified electrochemical charge storage properties of h-BN/rGO superlattice through the transition from n to p type semiconductor by fluorine doping, *Chemical Engineering Journal*, (2018). https://doi.org/10.1016/j.cej.2018.01.141.
95. S. Saha, W. Jang, N. C. Murmu, H. Koo, T. Kuila, Optimization of chemi-adsorption, EDLC, and redox capacitance through electro-precipitation synthesis of Fe_3O_4/NiO@rGO/h-BN for the development of hybrid supercapacitor, *Chemistry Select*, 4 (2019), 589–599. https://doi.org/10.1002/slct.201803611.
96. X. Jiao, T. You, F. Dai, P. Zhang, T. Li, F. Yu, L. Hu, L. Ding, L. Zhang, Z. Wen, Y. Wu, Hexagonal boron nitride nanosheet/carbon nanocomposite as a high-performance cathode material towards aqueous asymmetric supercapacitors, *Ceramics International*, 45 (2019), 4283–4289. https://doi.org/10.1016/j.ceramint.2018.11.101.
97. X. Zheng, G. Wang, F. Huang, H. Liu, C. Gong, S. Wen, Y. Hu, G. Zheng, D. Chen, Liquid phase exfoliated hexagonal boron nitride/graphene heterostructure based electrode toward asymmetric

supercapacitor application, *Frontiers in Chemistry*, 7 (2019), 544. https://doi.org/10.3389/fchem.2019.00544.

98. T. Yang, H. Liu, F. Bai, E. Wang, J. Chen, K. Chou, X. Hou, Supercapacitor electrode based on few-layer h-BNNSs/rGO composite for wide-temperature-range operation with robust stable cycling performance, *IntJ MinerMetallMater*, 27 (2020), 220–231. https://doi.org/10.1007/s12613-019-1910-x.

99. T. Gao, L. Gong, Z. Wang, Z. Yang, W. Pan, L. He, J. Zhang, E. Ou, Y. Xiong, W. Xu, Boron nitride/reduced graphene oxide nanocomposites as supercapacitors electrodes, *MaterialsLetters*, 159 (2015), 54–57. http://dx.doi.org/10.1016/j.matlet.2015.06.072.

100. S. Byun, J. H. Kim, S. H. Song, M. Lee, J-J. Park, G. Lee, S. H. Hong, D. Lee, Ordered, scalable heterostructure comprising boron nitride and graphene for high-performance flexible supercapacitors, *Chem. Mater.*, 28 (2016), 7750–7756. https://doi.org/10.1021/acs.chemmater.6b02947.

101. F. Song, Q. Chen, Y. Li, Y. Li, L. Zhang, High energy density supercapacitors based on porous $mSiO_2@Ni_3S_2/NiS_2$ promoted with boron nitride and carbon, *Chemical Engineering Journal*, 390 (2020), 124561. https://doi.org/10.1016/j.cej.2020.124561.

102. E. P. Gilshteyn, D. Amanbayev, A. S. Anisimov, T. Kallio, A. G. Nasibulin, All-nanotube stretchable supercapacitor with low equivalent series resistance, *Sci Rep.*, 7 (2017), 17449. https://doi.org/10.1038/s41598-017-17801-4.

103. S. Zheng, W. Lei, J. Qin, Z-S. Wu, F. Zhou, S. Wang, X. Shi, C. Sun, Y. Chen, X. Bao, All-solid-state high-energy planar asymmetric supercapacitors based on all-in-one monolithic film using boron nitride nanosheets as separator, *Energy Storage Materials*, 10 (2018), 24–31. http://dx.doi.org/10.1016/j.ensm.2017.08.002.

104. J. Hu, K. Xie, X. Liu, S. Guo, C. Shen, X. Liu, X. Li, J. Wang, B. Wei, Dramatically enhanced ion conductivity of gel polymer electrolyte for supercapacitor via h-BN nanosheets doping, *Electrochimica Acta*, 227 (2017), 455–461. http://dx.doi.org/10.1016/j.electacta.2017.01.045.

Chapter 3

Layered Double Hydroxides-Based Energy Storage Devices

Aruchamy Gowrisankar and Thangavelu Selvaraju

Department of Chemistry, Bharathiar University,
Coimbatore 641 046, Tamil Nadu, India

3.1 Introduction

Over the last few decades, all over the world, a rapid increase in global energy consumption has been noticed, due to the explosion of population. Upon the industrial revolution in the 19th century, the utilization rate of fossil fuels gained momentum for harnessing energy. But in the midst of the 20th century, environmentalists have understood the impact and consequences of fossil fuels, and the reserves are limited. Accordingly, the need for comprehensive research in the development of alternative energy storage devices was demanded to minimize the carbon footprint and retain the definite proportion of greenhouse gases in the atmosphere. Nowadays, the major innovations and attractive ideas in the domain of nanoscience and nanotechnology have decorated the front page of any leading scientific communications or in any dailies. In this modern era, many electrochemical energy

Layered Materials: Synthesis and Applications in Energy and Environmental Remediation
Edited by Alagarsamy Pandikumar, Kandasamy Jothivenkatachalam, and Subramanian Moscow

ISBN 978-981-5129-17-5 (Hardcover), 978-1-003-50831-1 (eBook)
www.jennystanford.com

storage systems such as batteries and supercapacitors have played an extensive role in the energy storage and conversion systems. Moreover, the criterion for energy storage devices must be balanced with the production of energy and power densities. The reduction in the size of a material to a nano regime has improved its electrical, mechanical, optical, and chemical properties from the bulk material [1–3]. As a result, the microscopic level of miniaturized batteries (MBs) and miniaturized supercapacitors (MSCs) can provide higher electrochemical activity. In general, supercapacitors (SCs) store energy either by faradaic reaction, where the energy is stored electrostatically, or by non-faradaic reaction, where redox reaction takes place between the electrodes. On the other hand, a battery stores energy via a reversible electrochemical reaction where charging and discharging occur slowly. In contrast to conventional batteries or SCs, MBs and MSCs are lightweight and portable, have minimum load leveling and superior rate capability, and are environmentally benign [4–6]. In addition, the electrochemical performance of the energy storage devices is dependent upon the materials that are used for electrodes [7–11]. Figure 3.1 represents the schematic representation of the mechanism involved in any electrochemical energy storage system. To date, different kinds of electrode materials, such as carbonaceous materials, conducting polymers, transition metal oxides/mixed metal oxides, transition metal chalcogenides (TMCs), layered double hydroxides (LDHs), and perovskite materials, have been employed as materials for conventional batteries and SCs [12–18]. Nevertheless, still, many efforts have been made in the development of an advanced electrode material with effective surface area. The materials under the nano regime are classified based on dimensions such as 0D, 1D, 2D, and 3D. Among different dimensional materials, 2D electrode materials are very attractive due to their (a) large surface area, (b) easy functionalization, (c) high mechanical flexibility, (d) thickness control, (e) fast transportation of electrolyte ions between two layers, and (f) tunable band gap for both conventional and miniaturized batteries and supercapacitors [19, 20]. Thereby, 2D materials such as graphene, graphitic carbon nitride, TMC, and LDH are known as established electrode materials in any electrochemical energy storage [13–18, 21]. The 2D materials are generally used in the form of nanosheets, which consist of covalently bonded monolayers in nanometer size. Recent advancement has been made for 2D architecture-based microelectronic devices [22]. On the other hand, many efforts have to be introduced in 2D materials such as MBs or MSCs, mostly due to their design and microfabrication techniques remaining undeveloped [23]. There are many simple fabrication techniques, such as direct writing, gravure printing, electrochemical deposition, and laser scribing, used to design and ease out fabrication of MBs and MSCs [24, 25]. In addition, the intercalation of a variety of anions into 2D materials such as LDHs offers higher electrochemical activity via the fast electron transfer process between two interlayers. Based on the merits, it is possible to reveal that LDHs could be one of the suitable 2D electrode materials for electrochemical energy storage devices.

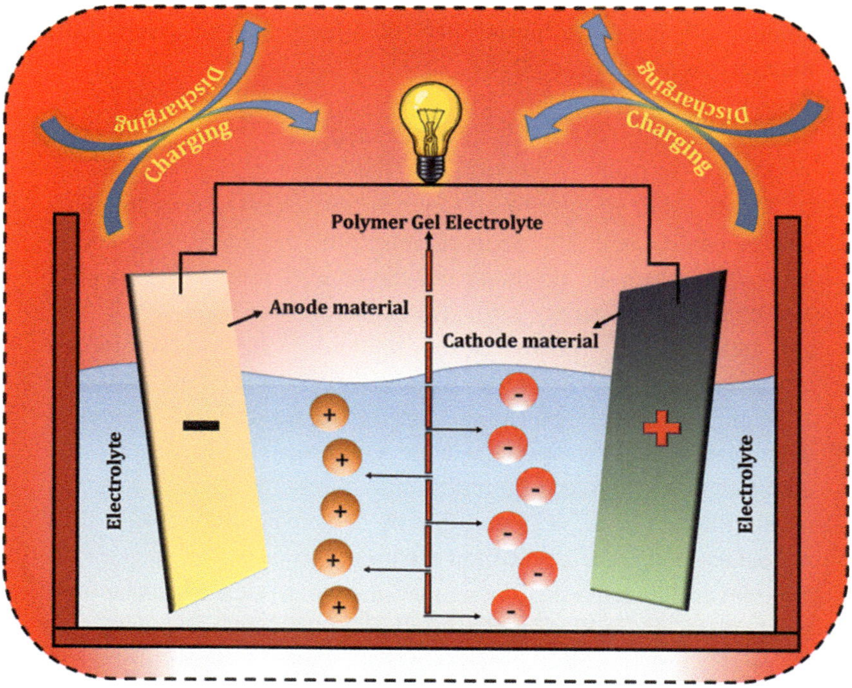

Figure 3.1 Schematic representation of the electrochemical energy storage mechanism.

3.2 Layered Double Hydroxide: An Overview

Contemporary LDHs have significant research interest, due to their wide range of applications such as catalysts, biomaterials, polymer additives, environmental materials, thin films, fire retardant additives, anti-corrosion coatings, and controlled drug release in medications [26]. LDH's chemical composition is represented by its general formula of $[M_{(1-x)}{}^{2+}M_x{}^{3+}(OH)_2]_{layer}$ $[A_{x/n}{}^{n-}\cdot zH_2O]_{interlayer}$, where M^{2+} and M^{3+} are divalent (M^{2+} = Mg, Zn, Ni, Ga, Mn, Ca, In, V, Co, and Cu) and trivalent (M^{3+} = Al, Fe, and Cr) metal cations occupying the octahedral interstices of the layers, while A^{n-} represents the charge balancing organic or inorganic anions. Moreover, any organic or inorganic anions can be introduced between hydroxide layers by ion exchange method or co-precipitation method. The x value signifies the molar ratio of M^{2+} or M^{3+} metal ions. Commonly used anions in LDHs include halides, oxoanions, oxometalates, polyoxometalates, and organic anions [27, 28]. The 'preference' of affinity for anions in LDH materials is as follows:

$$CO_3^{2-} > SO_4^{2-} > OH^- > F^- > Cl^- > Br^- > NO_3^- > I^-$$

LDHs occur both naturally or can be synthesized at low cost and by environmentally clean chemistry methodologies in several ways. The synthetic methods include pulsed laser ablation [29], microwave irradiation [30], hydrothermal method [31], sol-gel method [32], reconstruction method [33], co-precipitation method [28], and urea method [34]. Using any of these methods, a large number of possible LDH materials can be produced with high porous properties. Notably, the methodology for the direct synthesis in the generation of monolayer LDHs remains impossible; whereas, it is possible by chemical exfoliation of LDH materials under ultrasonication technique to produce effective 2D monolayer nanosheets [35]. Figure 3.2 illustrates the formation mechanism of LDH materials. Initially, the nucleation process of metal precursors takes place followed by crystal growth that occurs homogenously where sodium hydroxide acts as a precipitating agent in the formation of aggregated LDH crystals, which exfoliated by ultrasonication could generate 2D nanosheets.

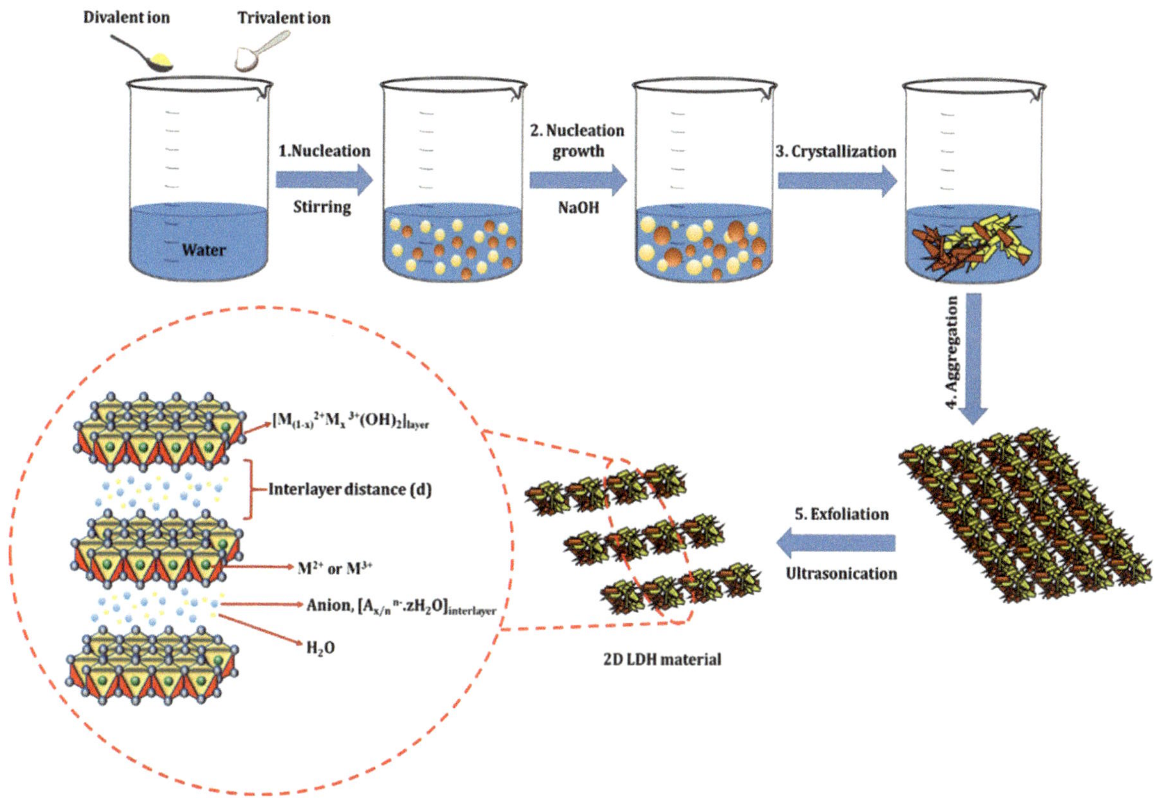

Figure 3.2 Schematic representation of the formation mechanism of 2D LDH materials with detailed structure.

3.3 LDH-Based Battery Devices

The limitations in the MBs are constrained by their short lifecycle and continuous structural changes that occur during the lifecycle process. Many efforts have been undertaken by scientific researchers to overcome the circumstances. According to preceding reports, nickel-bismuth (Ni//Bi)-based rechargeable aqueous batteries are receiving considerable attention, owing to their high operating potential window and excellent storage capacity [36]. Meanwhile, TMOs or mixed TMOs, such as MnO_2, NiO_2, SnO_2, Co_3O_4, $Ni(OH)_2$, etc., were investigated as the cathode materials [10, 24, 28, 37]. Recently, J. Wang and S. Yang et al. have developed a Ni//Bi battery using $Ni(OH)_2$/graphene sheets as positive electrode and Bi_2O_3 rods/graphene sheets as negative electrode to deliver a high discharge capacity of 102 mAh g^{-1} at 1 C and good rate capability [36]. Further, L. Mai and W. Luo et al. have assembled a high-energy-density Ni//Bi battery, where Bi and $Ni_{1-x}Co_x(OH)_2$ act as anode and cathode materials, respectively [38]. Briefly, the Ni//Bi battery was constructed through a micro-electrochemical system (Fig. 3.3 A). Initially, PR1-9000A photoresist film was coated on the Si/SiO_2 wafer by the spin coating process. Afterward, four pairs of coplanar interdigital micro-patterns of photoresists were obtained by photolithographic technique on the silicon wafer. Then, an interdigital micro-pattern of the Cr/Au current collector was deposited by physical vapor deposition. Afterward, Bi film was electrochemically deposited at the Cr/Au current collector using the three-electrode system and electrochemical workstation. Besides, Bi-200 samples were also prepared by calcination at 200 °C under a nitrogen atmosphere for 1 h. On the other side, $Ni_{1-x}Co_x(OH)_2$ was synthesized and deposited on the Cr/Au current collector. Finally, PVA/KOH solid gel solution was added drop by drop on the micro-electrode to assemble an alkaline planar micro-battery. Figure 3.3 B shows the cyclic voltammetry (CV) curves of Bi-200 film with a potential window from −1.1 to 0 V; whereas, the potential window is set between 0 and 0.5 V for $Ni_{1-x}Co_x(OH)_2$ at a scan rate of 20 mV s^{-1} in the three-electrode system. Subsequently, the electrochemical performance of the assembled Ni//Bi-based MB was also studied in the two-electrode system.

The CV curves of Ni//Bi MB at different scan rates with the potential from 0 to 1.6 V were recorded (Fig. 3.3 C). It is seen that the redox peak shift to higher voltage indicates the effective electron transfer reaction in the Ni//Bi MB. Further, the galvanostatic charge-discharge (GCD) profile shows the charge storage capacity of the Ni//Bi MB at different applied current densities from 1 to 5 mA cm^{-2} (Fig. 3.3 D). Ni//Bi MB device displays an increase of potential with a rise in the potential window. From Fig. 3.3 E, the specific capacity of the different current densities was plotted where the higher areal specific capacity was observed as 55.40 mC cm^{-2} at a low current density of 1 mA cm^{-2}. Moreover, the areal specific capacity of 44.03 mC cm^{-2} was observed even at a higher current density of 5 mA cm^{-2} to imply the high-rate capability and capacitance retention of Ni//Bi MB. Further, it demonstrates the capacity retention of 78.51% with the high-rate capability during an increase in the current density from 1 to 5 mA cm^{-2}. In addition, the cycling performance of the electrode material has played a significant role in the practical applications of energy storage devices. As shown in Fig. 3.3 F, the capacitance retention of 84.44% was detected, which confirms the good cycling performance of Ni//Bi MB. Here, the cycling test was conducted at 10 mA cm^{-2} current density. The Ragone plot (Fig. 3.3 G) shows that the Ni//Bi MB has achieved a maximum energy density of 12.3 μWh cm^{-2} at a power density of 0.799 mW cm^{-2}. Nevertheless, it maintains a remarkable energy density of 9.5 μWh cm^{-2} even at a higher power density of 3.98 mW cm^{-2}, suggesting the superior charge-discharge capability of Ni//Bi MBs. The Ragone plot results are comparable with the previously reported values such as MoS_2/reduced graphene oxide-carbon nanotube, MnO_2//Fe_2O_3 MSC, MnO_2-based MSC, and 3D graphene pellet-MSC, signifying the retention of higher energy density of Ni// Bi [39–42]. Therefore, a planar $Ni_{1-x}Co_x(OH)_2$ and Bi-200-based Ni//Bi MB could be a promising electrode material for practical applications.

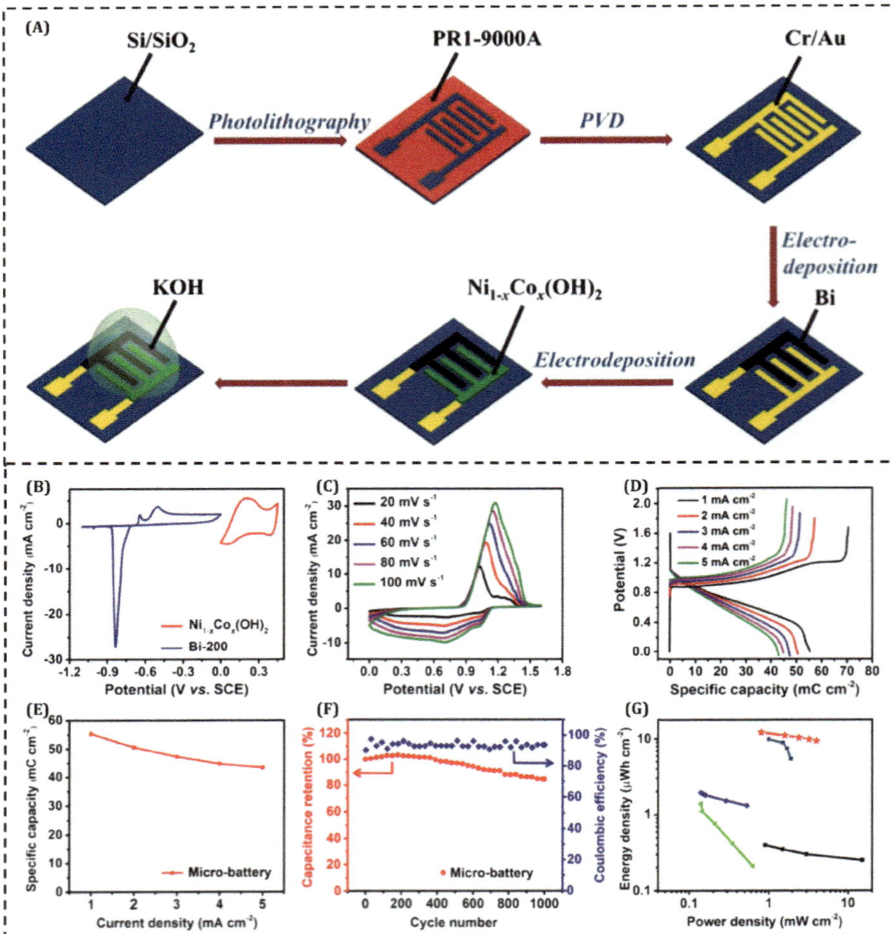

Figure 3.3 (A) Schematic illustration of the microfabrication process of the planar alkaline Ni//Bi MB. (B) CV curves of the Bi-200 film and $Ni_{1-x}Co_x(OH)_2$ loaded electrode in the three-electrode system. (C) CV curves at different scan rates of Ni//Bi MB. (D) GCD curves at various current densities. (E) Relationship between specific capacity and current density. (F) Cycling performance at 10 mA cm^{-2} current density. (G) Ragone plot.

In recent days, flexible and wearable electronic devices have received tremendous attention due to their superior physical stability. The main objective of flexible energy storage devices is the incorporation of flexible electrode materials or a combination of 2D and 3D materials on the substrate [43]. L. He and L. Mai research group has introduced a novel wearable electrode material using Co and Zn, resulting in the development of alkaline Co-Zn MB [44]. As shown in Fig. 3.4 A, the fabrication process was initially started with the coating of Ni-layer on the pristine textile (NT) by the electroless plating process. The formed NT shows higher electrical conductivity with a low resistance of 4.7 Ω. Afterward, the hierarchical $Co(OH)_2$@NiCo-LDH@NT(CH@NC-LDH@NT) core-shell electrode as a cathode material was fabricated via the hydrothermal method followed by the electrodeposition process. Furthermore, the laser engraving technique was introduced to cut the CH@NC-LDH@NT electrode material in a micro-sized pattern. Meanwhile, the 3D porous Zn dendrite was electrodeposited on the carbon cloth (Zn@CC) under identical conditions to act as an anode material. Finally, the obtained micro-anode and micro-cathode were properly placed on the suitable polydimethylsiloxane (PDMS) based micro-trench, followed by dropping the polymer gel electrolyte. Obviously, PDMS gel electrolyte is an insulating material that separates the anode and cathode without any structural degradation. Here, the formed device shows a geometric area of 0.96 cm^2. Figure 3.4 B shows the CV curves of CH@NC-LDH@NT and Zn@CC electrode at 5 mV s^{-1} in the three-electrode system to confirm the presence of two pairs of redox peaks at 0.17/0.41 V and 0.26/0.46 V that are attributed to the existence of Ni^{2+}/Ni^{3+} and $Co^{2+}/Co^{3+}/Co^{4+}$ at CH@NC-LDH@NT electrode. Besides, a pair of redox peaks were obtained at −1.25 and −1.52 V for the Zn@CC electrode. Moreover, a large potential difference of 1.65 V was observed between two redox peaks indicating the high output voltage of assembled Co-Zn MB. From Fig. 3.4 C, the CV curves exhibit no obvious changes at different scan rates, which leads to the good electrochemical stability of Co-Zn MB. The plausible overall electrochemical reaction is illustrated as follows:

$$Zn + CoOOH + NiOOH + 2KOH + 2H_2O \leftrightarrow K_2[Zn(OH_4)] + Ni(OH)_2 + Co(OH)_2$$

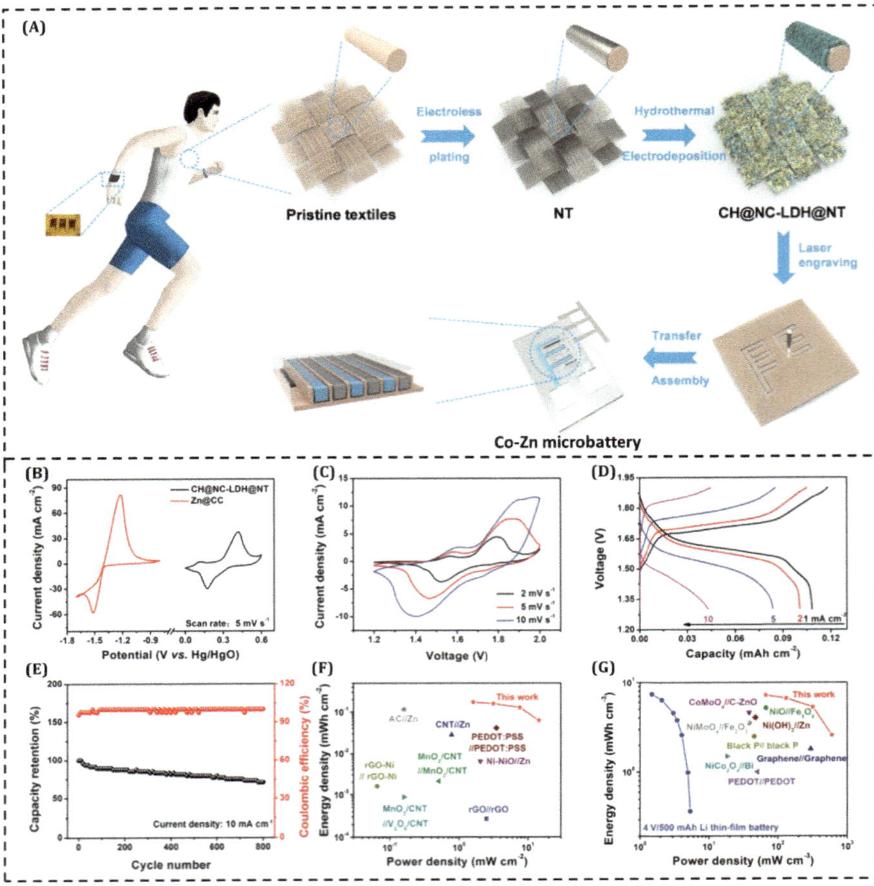

Figure 3.4 (A) Schematic representation of the microfabrication of alkaline Co-Zn MB. (B) CV curves of the CH@NC-LDH@NT and Zn@CC at 5 mV s^{-1} in the three-electrode system. (C) & (D) CV and GCD curves at various scan rates and current densities, respectively. (E) Cycling life and its Coulombic efficiency. (F) & (G) show Ragone plots of Co-Zn MB.

Then, the GCD curves were recorded at different current densities such as 1, 2, 5, and 10 mA cm^{-2}, where the charging and discharging plateaus are coherent with CV results (Fig. 3.4 D). The calculated areal capacities of 108.3, 100.6, 83.6, and 43.1 μAh cm^{-2} at different current densities have indicated the excellent electrochemical capacity of Co-Zn MB. The cycle life and Coulombic efficiency of alkaline batteries were studied in constant GCD curves recorded at 10 mA cm^{-2} current density. Figure 3.4 E exhibits outstanding cyclic stability and retains 71% of initial capacity with 100% Coulombic efficiency over 800 cycles. The morphological changes that occur in the cycling test have decreased the cyclic stability of Co-Zn MB. The corresponding Ragone plot related to the energy and power densities was calculated at various current densities and compared with the reported literature. As can be seen in Fig. 3.4 F, Co-Zn MB exhibits a real energy density of 0.17 mWh cm^{-2} with the maximum areal power density of 14.4 mW cm^{-2}, which is higher than the most reported literature such as rGO//rGO (0.00051 mWh cm^{-2} and 2.51 mW cm^{-2}) [45], rGO-Ni//rGO-Ni (0.00158 mWh cm^{-2} and 1.86 mW cm^{-2}) [46], CNT//Zn (0.028 mWh cm^{-2} and 7.8 mW cm^{-2}) [47], Ni-NiO//Zn (0.0066 mWh cm^{-2} and 20.2 mW cm^{-2}) [48], PEDOT:PSS//PEDOT:PSS (0.041 mWh cm^{-2} and 3.52 mW cm^{-2}) [49], MnO_2/CNT//MnO_2/CNT (0.02107 mWh cm^{-2} and 0.51 mW cm^{-2}) [50], AC//Zn (0.1154 mWh cm^{-2} and 3.9 mW cm^{-2}) [51], MnO_2/CNT//V_2O_5/CNT (0.00088 mWh cm^{-2} and 0.63 mW cm^{-2}) [52], and $Cu(OH)_2$@FeOOH/Cu (0.01807 mWh cm^{-2} and 0.73 mW cm^{-2}) [53]. In addition, the volume energy and power densities were also calculated. The Co-Zn MB delivers a higher volume energy density of 7.23 mWh cm^{-3} with 599.9 mW cm^{-3}, which is higher than those of the reported values such as Ni-NiO//Zn (0.67 mWh cm^{-3} and 220 mW cm^{-3}) [48], graphene//graphene (1.81 mWh cm^{-3} and 297 mW cm^{-3}) [54], $CoMoO_4$//C-ZnO (4.6 mWh cm^{-3} and 420 mW cm^{-3}) [55], NiO//C-ZnO (7.76 mWh cm^{-3} and 210 mW cm^{-3}) [56], $Ni(OH)_2$@Ni//Zn@Ni (4.05 mWh $cm^{-3}$and 425 mW cm^{-3}) [57], 4V/500 μAh Li thin-film battery (7.4 mWh cm^{-3} and 5.4 mW cm^{-3}) [58], PEDOT//PEDOT (1 mWh cm^{-3} and 1000 mW cm^{-3}) [59], NiO//Fe_3O_4 (5.2 mWh cm^{-3} and 640 mW cm^{-3}) [60], and $NiMoO_4$//Fe_2O_3 (0.68 mWh cm^{-3} and 330.62 mW cm^{-3}) [61]. Therefore, based on the above results and due to the high electrochemical performance, it is possible to promote Co-Zn MB as a promising candidate for miniaturized batteries.

In order to check the physical flexibility of Co-Zn MB, a series of mechanical tests were conducted with different bending angles. The original position of the micro-device without applying the force is depicted in Fig. 3.5 A. Further, lateral and vertical bending

(Fig. 3.5 B and C) of the micro-device has delivered the same relative position of the electrodes to maintain similar specific capacity. From Fig. 3.5 D, the micro-device can be directly attached to the human arm, indicating the easy integration of the wearable Co-Zn MB. In addition, the flexibility of Co-Zn MB was further tested under various bending angles (θ = 0°, 45°, 90°, and 180°) with constant bending radius (R = 10 mm) and device length (L = 30 mm). As a result, no obvious changes were observed in the specific capacity, which confirmed that Co-Zn MB has retained an outstanding physical stability structure. Besides, the cyclic stability test was conducted (after 200 cycles) by bending the micro-device at a 90° bending angle with a similar bending radius (Fig. 3.5 E). Interestingly, excellent capacity retention of 93.6% was obtained after 200 times bending the Co-Zn MB (Fig. 3.5 F). As a result, Co-Zn MB has proved its unique structural configuration with high flexibility and wearability.

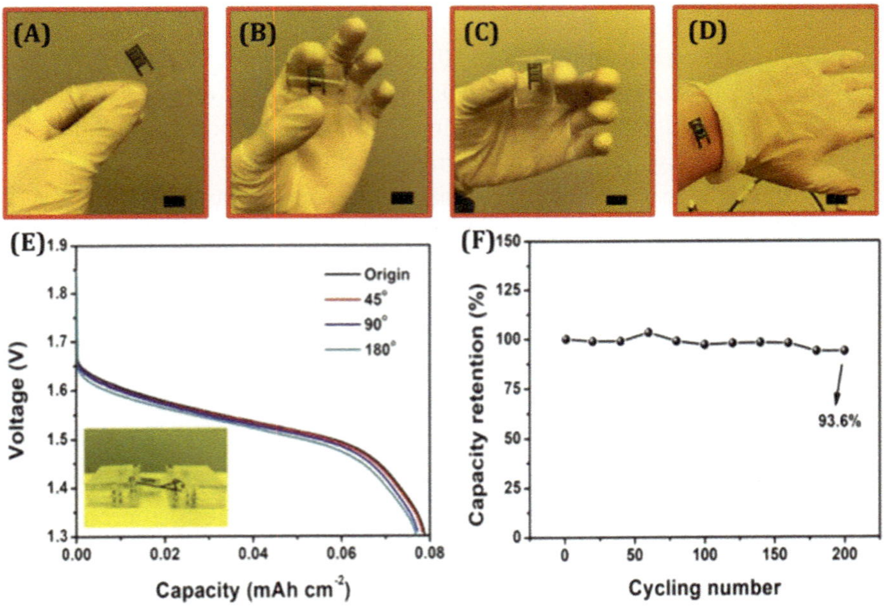

Figure 3.5 Optical photographs of the mechanical performances of Co-Zn MB, (A) flat or original position, (B) lateral bending, (C) vertical bending, and (D) attached on skin. (E) GCD curves of the Co-Zn MB under different bending angles (0–180°). Bending radius, R = 10 mm and device length, L = 30 mm, and (F) capacity retention test at 90° bending angle over 200 cycles.

Consecutively, lithium-sulfur (Li-S) batteries possess high theoretical specific capacity (1675 mAh kg^{-1}) and energy density (2600 Wh kg^{-1}) and are termed next-generation electrochemical energy storage devices [62]. However, the implementation of Li-S batteries still suffers, due to the shuttle effect caused by the dissolved lithium polysulfides (LiPSs) during cyclic study. Therefore, Li-S batteries possess low-rate capacity, low-Coulombic efficiency, and loss of active sulfur. J. Wang and S. Deng research group has designed a new sulfur host cathode to physically attach with nanopores present in the LiPSs. Meanwhile, during the long-term cyclic stability studies, the polar LiPSs have diffused outward, due to their weak interaction with non-polar carbon nanopores. In addition, the poor kinetics of LiPSs reduction and transformation lead to the precipitation of Li_2S_2/Li_2S at the cathode/anode surface. Consequently, various metal oxides, conducting polymers, and metal sulfides were introduced to strengthen the interaction between polar LiPSs and non-polar carbon hosts. Therefore, a novel synthetic route and fabrication technique is required for an enhanced electrochemical performance of LiPSs. Additionally, Co- and Ni-based metal sulfides/nitrides have been proposed as efficient electrocatalysts for the transformation of LiPSs to Li_2S_2/Li_2S [63].

Dong et al. have facilitated the transformation of LiPSs to Li_2S_2/Li_2S where nitrogen heteroatom has accelerated the oxidation of LiPSs to S_8 using Co [64]. In addition, Wu et al. demonstrated Co nanoparticles and nitrogen heteroatoms encapsulated graphitic carbon network-based catalysts to easily solubilize the LiPSs and performed an enhanced kinetic redox reaction. On the contrary, the preparation of catalysts is a time-consuming process [65]. Therefore, J. Wang and S. Deng research group has prepared a novel and low-cost hollow cobalt-embedded and nitrogen-doped carbonaceous micro-flower (H-Co-NCM) via in situ transformation of metanilic anions intercalated Co-Al layered double hydroxides (CoAl-M-LDHs) [63]. The detailed fabrication technique is shown in Scheme 3.1. Initially, the H-Co-NCM was prepared by calcination of CoAl-M-LDHs at 700 °C to form nitrogen-doped carbon skeletons (c-LDHs). Further, the acid treatment was carried out with HCl to obtain homogeneous particle size and ordered morphology, where sulfur atoms are encapsulated into H-Co-NCM using CS_2 solvent to form S@H-Co-NCM composites.

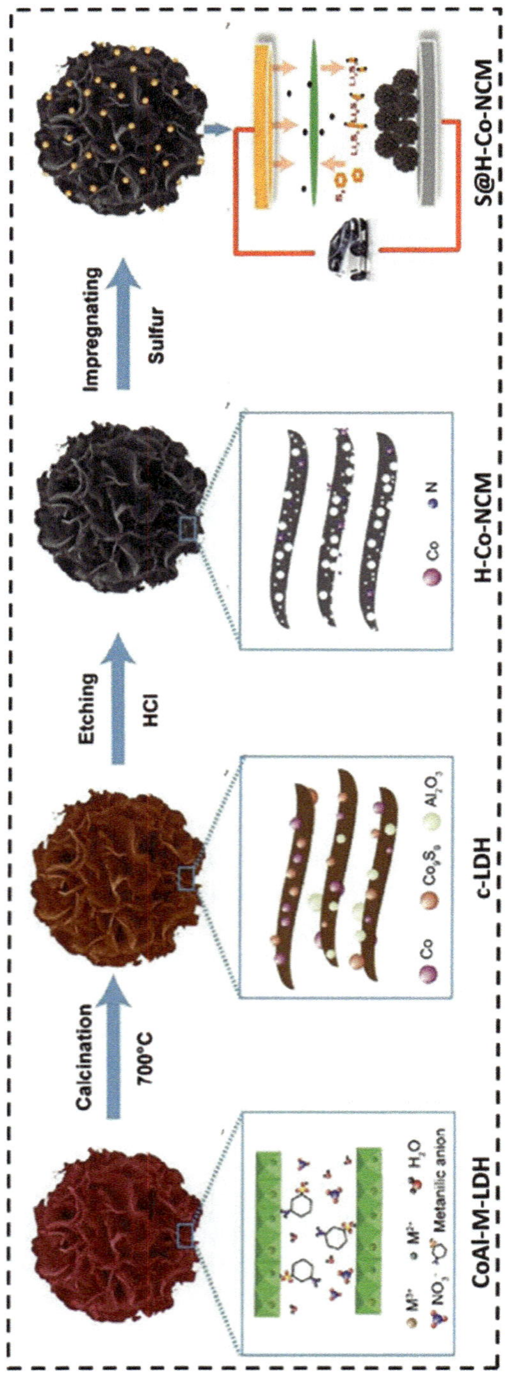

Scheme 3.1 Schematic illustration of the synthesis procedure of an S@H-Co-NCM electrode material.

SEM and TEM images of CoAl-M-LDH, c-LDH, and H-Co-NCM (Fig. 3.6) have inferred a uniform distribution of 3D flower-like morphology for CoAl-M-LDH (Fig. 3.6 A), with the average diameter of ~2 μm. The assembly of a 3D flower-like structure was obtained by the organization of multiple nanosheets, which are stabilized by the structural integrity during the sulfur impregnation and cycling process. During the calcination process of CoAl-M-LDH at 700 °C, metanilic acids are converted into an N-doped carbon framework. Meanwhile, the LDH layers are transformed into Co and Al nanoparticles, where the nanoparticles are uniformly dispersed on the petals of micro-flower with a thickness of about 30 nm (Fig. 3.6 B, E, H). Subsequently, the excess Co and Al nanoparticles were etched away by HCl treatment, which resulted in the formation of the H-Co-NCM nanostructure (Fig. 3.6 C, F, I). As a result, the nanostructure provides numerous channels for fast electrolyte insertion/de-insertion. Further, the unique hollow nanostructure was obtained by encapsulation of sulfur particles into the H-Co-NCM framework (1–20 nm thickness) (S@H-Co-NCM) to block the diffusion pathways of LiPSs.

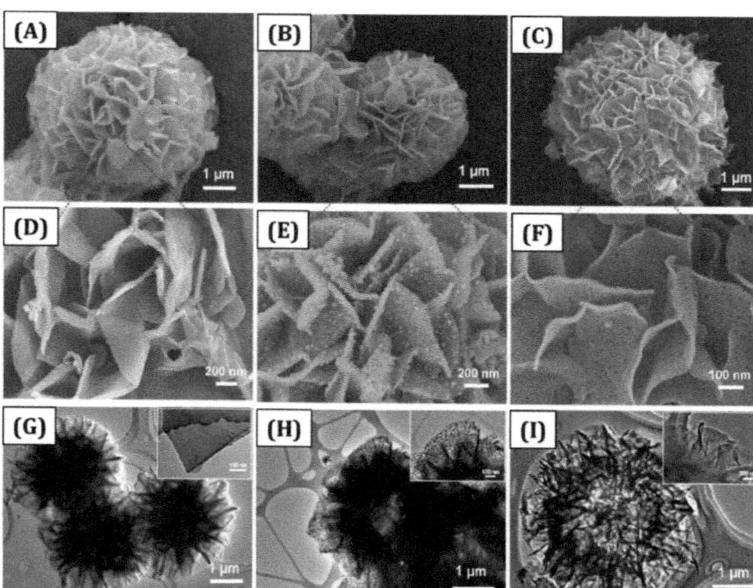

Figure 3.6 SEM images of CoAl-M-LDH (A, D), c-LDH (B, E), and H-Co-NCM (C, F). TEM images of CoAl-M-LDH (G), c-LDH (H), and H-Co-NCM (I).

Besides, the electrochemical performance of S@H-Co-NCM was performed by investigating CV and EIS profiles. Here, acetylene black and sulfur composite (S@CB) acted as a control. As shown in Fig. 3.7 A, the two sharp reversible redox peaks were observed. The two cathodic or anodic peaks were denoted as I_{C1}, I_{C2} and I_{A1}, I_{A2}, respectively. The two signature redox peaks corresponding to the reduction of sulfur to Li_2S_2/Li_2S and reversible oxidation of Li_2S_2/Li_2S to S_8 were observed. The CV curves of the S@H-Co-NCM electrode show a sharper redox peak with a superior positive shift. Notably, both the anodic peaks appeared at the same position, indicating that the Co-doped and N-doped functionalities have decreased the electronic polarization to enhance the polysulfide redox reactions. Further, the EIS spectrum was recorded as shown in Fig. 3.7 B. The calculated charge transfer resistance (R_{ct}) for S@H-Co-NCM (62.11 Ω) is lower than the R_{ct} of S@CB (80.09 Ω), which suggests the occurrence of faster reaction kinetics in Co- and N-doped functionalities. Besides, the different GCD curve at various current rates was shown in Fig. 3.7 C, where a meager variation in the voltage difference on the GCD curves of S@H-Co-NCM (152 mV) was observed compared to S@CB (156 mV). During the enhancement in the current rate from 0.1 C to 1 C, the S@H-Co-NCM cathode has displayed two discharge plateaus; whereas, S@CB has shown a plain discharge curve.

It is due to the severe polarization effect on the surface of the S@CB cathode material; whereas, the kinetics of the polysulfide conversion has improved when increasing the current rate at the S@H-Co-NCM cathode. Upon increasing the current densities from 0.1, 0.2, 0.5, 1, and 2 C, the S@H-Co-NCM cathode delivers an outstanding storage capacity of 1374, 957, 841, 738, and 611 mAh g^{-1}, respectively. After switching back the current densities from 2, 1, 0.5, 0.2, and 0.1 C, the obtained corresponding discharge capacity was recovered as 768, 797, 813, and 829 mAh g^{-1} (Fig. 3.7 D). The S@H-Co-NCM electrode material has higher structural stability during the cycling test. Meanwhile, S@CB exhibits capacity decay while increasing the current density. Moreover, the long-term cycling test was conducted (Fig. 3.7 E) at a 0.5 C current rate. After 500 cycles, the capacity decay of 0.069% was observed with 100% Coulombic efficiency

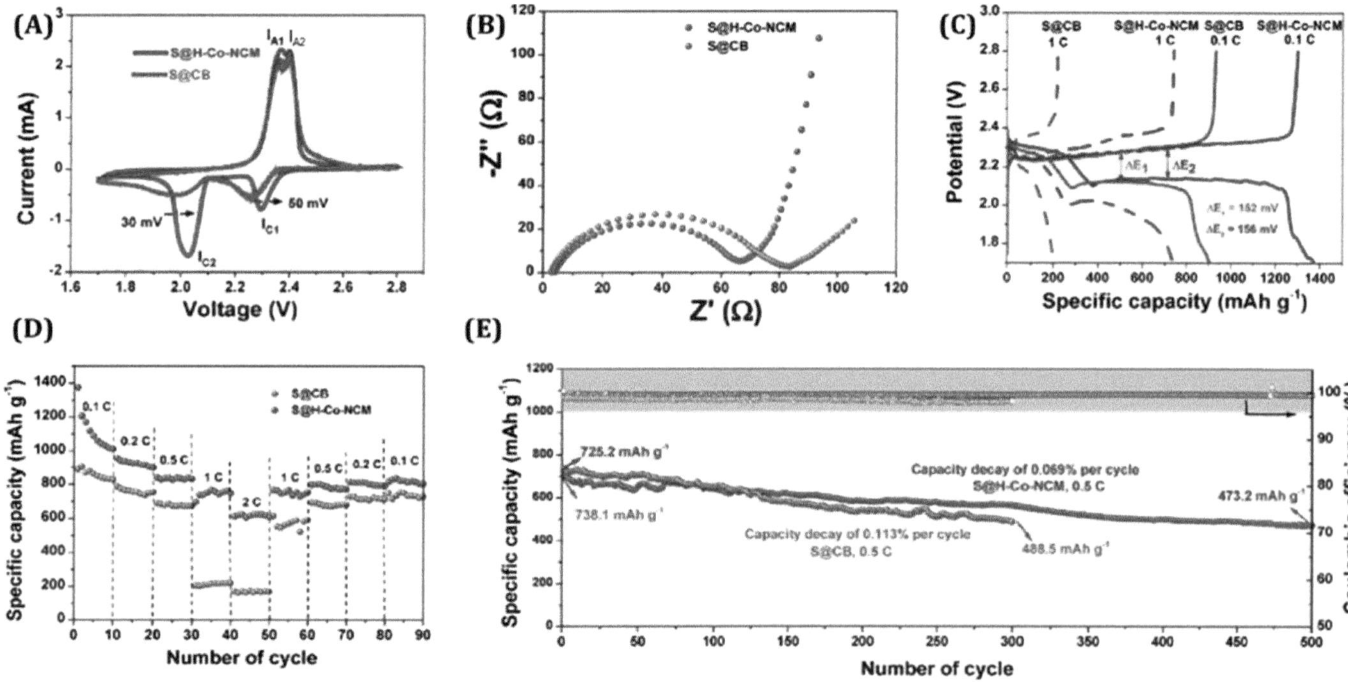

Figure 3.7 Electrochemical performances of S@H-Co-NCM and S@CB cathodes. CV curves (A), EIS spectrum (B), GCD curves at 0.1 and 1 C current density (C), rate performance from 0.1 to 2 C current density (D), and cycling test at 0.5 C current density (E).

for S@H-Co-NCM cathode material; whereas, S@CB cathode material showed 0.113% capacity decay after 300 cycles. As a result, S@H-Co-NCM cathode material exhibits excellent electrochemical performance, which is attributed to the strong chemisorption and fast electrocatalytic effects provided by the Co and N functionalities. This study has reflected the exploration of novel cathode materials with high specific storage capacity for the fabrication of Li-S batteries.

3.4 LDH-Based Supercapacitors

Supercapacitors are efficient electrochemical energy storage devices by reversible adsorption and desorption of electrolyte ions at the interface between electrolyte and active electrode material [11]. Recently, low-dimensional bimetal sulfides such as $NiCo_2S_4$ have been frequently used to fabricate asymmetric supercapacitors by the multiple valences to promote fast reversible redox reactions [66]. Interestingly, S. Deng and J. Wang research group found a highly conductive $NiCo_2S_4$ nanotube to provide a high-speed pathway for the transportation of electrolyte ions and also to support the growth of glucose intercalated NiMn-LDH (NiMn-G-LDH) nanosheets [67]. Moreover, the intercalation of glucose into the NiMn-LDH layer has significantly increased the interlayer distance and boosted the ion transport rate. NiMn-G-LDH layers were orderly arranged on the $NiCo_2S_4$ nanotube and subsequently coated on carbon fiber cloth (NiMn-G-LDH@$NiCo_2S_4$@CFC) as a free-standing substrate. Scheme 3.2 shows the detailed procedure for the formation of NiMn-G-LDH@$NiCo_2S_4$@CFC core-shell heterostructure. Firstly, NiCo-LDH nanoneedle was grown on a CFC substrate by the hydrothermal method. Secondly, CO^{3-} and OH^- ions were replaced with S^{2-} ions by adding Na_2S solution. Subsequently, the prepared NiMn-G-LDH was uniformly wrapped on the $NiCo_2S_4$@CFC core electrode material.

Furthermore, the morphologies of the as-prepared core-shell heterostructure were examined by the FESEM technique (Fig. 3.8). As shown in Fig. 3.8 A, NiCo-OH LDH was uniformly coated on the CFC substrate. The vertically aligned NiCo-OH LDH nanoneedles have 1 μm length and ~50 nm diameter (Fig. 3.8 A–C).

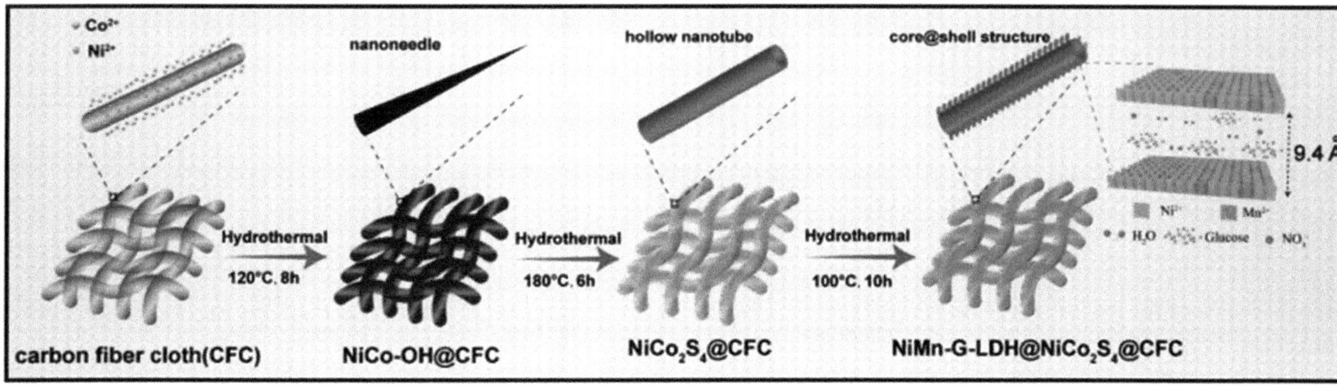

Scheme 3.2 Schematic illustration of the formation of NiMn-G-LDH@$NiCo_2S_4$@CFC.

Figure 3.8 FESEM images of NiCo-OH@CFC (A–C), $NiCo_2S_4$@CFC (D–F), and NiMn-G-LDH@$NiCo_2S_4$@CFC (G-I).

After adding Na_2S solution, the coated material turned to a black color indicating the formation of $NiCo_2S_4$ nanotubes on CFC (Fig. 3.8 D–F). The conversion of nanoneedles to nanotubes is due to the Kirkendall effect, where the unequal diffusion of Co^{2+} and S^{2-} ions are developed in the inner void space. After depositing the NiMn-G-LDH, the black color changes into a light gray color, indicating the successful formation of NiMn-G-LDH@ $NiCo_2S_4$@CFC (Fig. 3.8 G–I). Unlike 2D vertically aligned LDH nanosheets grown on CFC, the interconnected hollow nanotubes wrapped by NiMn-G-LDH nanosheets have facilitated more active sites for fast transport kinetics of the electrolyte ions. Importantly, the NiMn-G-LDH shell has acted as a protective layer to prevent the structural dwindling on the cycling test.

In order to fabricate the flexible all-solid-state asymmetric supercapacitor (FSAS), NiMn-G-LDH@$NiCo_2S_4$@CFC acted as positive and activated carbon as negative electrodes separated by solid gel PVA/KOH as a separator. The schematic diagram of the assembled FSAS device is shown in Fig. 3.9 A. Moreover, the CV profile of the NiMn-G-LDH@$NiCo_2S_4$@CFC//AC device under different potential windows such as 1.2, 1.3, 1.4, and 1.5 V at 5 mV s^{-1} scan rate is shown in Fig. 3.9 B. It clearly shows that the CV curves of NiMn-G-LDH@$NiCo_2S_4$@CFC//AC device are highly stable even at a 1.5 V potential window. Further, the CV curves at different scan rates (from 5 to 50 mV s^{-1}) with the potential window ranging from 0 to 1.5 V were shown (Fig. 3.9 C), which retained the shape of CV curve even at a higher scan rate, suggesting the good rate capability. From Fig. 3.9 D, the symmetrical GCD curves under different current densities (from 0.5 to 10 A g^{-1}) were obtained to show the high-Coulombic efficiency of the FSAS device. Further, the higher specific capacitance of 93.0 F g^{-1} was attained at 0.5 A g^{-1} to maintain 62.6 F g^{-1} at 10 A g^{-1} current density (Fig. 3.9 E). Moreover, the long-term cycling test was conducted at 10 A g^{-1} current density and exhibited 86.4% initial capacitance even after 10,000 cycles (Fig. 3.9 F). In addition, the two FSAS devices were connected in series and evaluated by GCD curves. The potential window is extended from 1.5 to 3 V for two devices (Fig. 3.9 G). In practical applications, flexibility is an important property of any energy storage device. The mechanical stability test was conducted under different bending angles such as 120° and 160°. The obtained CV curves almost overlap each other, indicating that the device has excellent flexibility and structural stability (Fig. 3.9 H). On the other hand, the intrinsic performance of the FSAS device was further measured by calculating the power and energy density, and the results were compared with previously reported core-shell materials. Evidently, the FSAS device possesses an energy density of 60.3 Wh kg^{-1} at a power density of 375 W kg^{-1} and retained an energy density of 19.6 Wh kg^{-1} at a higher power density of 7506 W kg^{-1}. Noticeably, the FSAS device has delivered the maximum energy and power density than NiCo-LDH/NiCoP@NiMn-LDH (42.2 Wh kg^{-1} at 750 W kg^{-1}) [68], $NiCo_2S_4$@$Ni(OH)_2$@PPy (34.7 Wh kg^{-1} at 120 W kg^{-1}) [69], $NiCo_2S$ nanopetals//AC (35.6 Wh kg^{-1} at 819.5 W kg^{-1}) [70], (Ni, Co)Se_2/NiCo-LDH //AC (39.0 Wh kg^{-1} at 1650 W kg^{-1}) [71], and CoS_x/Ni-Co LDH//AC (35.8 Wh kg^{-1} at 800 W kg^{-1}) [72]. Thus, the remarkable electrochemical performance of FSAS the device is mainly attributed to the unique core-shell nanostructure of the electrode material. More interestingly, the hollow structure of the $NiCo_2S_4$ provides a high-speed pathway for the electrolyte ions. The electrochemical reaction mechanism of the reversible redox reaction is given as:

$$NiCo_2S_4 + 2OH^- \leftrightarrow NiCo_{4-2x}\,OH + 2CoS_xOH + 2e^-$$

$$CoS_xOH + OH^- \leftrightarrow CoS_xO + H_2O + 2e^-$$

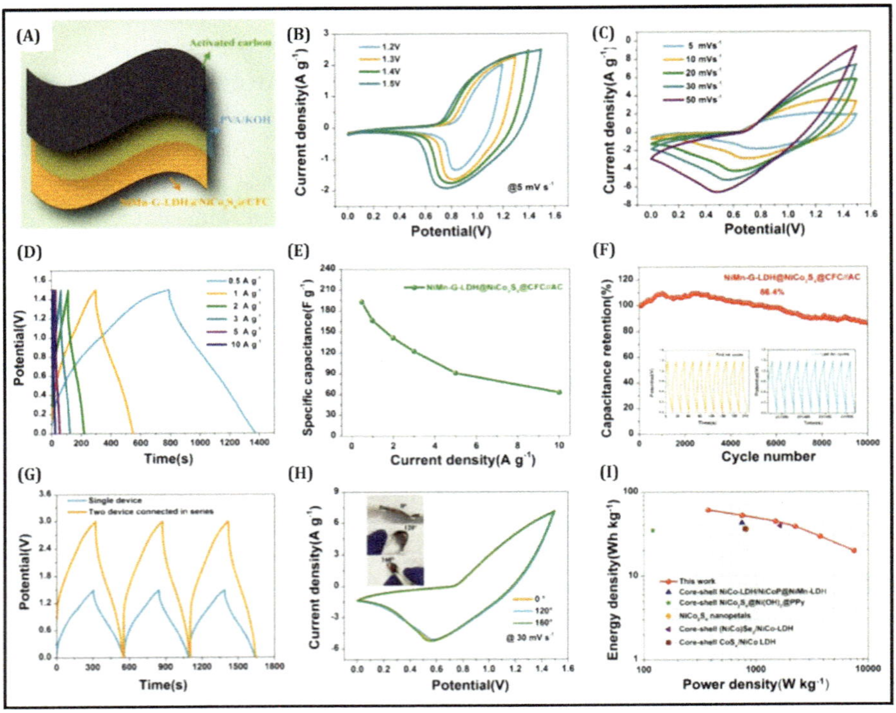

Figure 3.9 Electrochemical performance of the FSAS device. Schematic diagram of the device (A), CV profile under various potential windows at 5 mV s^{-1} scan rate (B), CV curves (C), and GCD curves (D) at different scan rates and current densities, respectively. Plot between specific capacitance and current density (E), the long-term cyclic stability study (F), GCD curves for single and two devices connected in series (G), CV curves under various bending angles at 30 mV s^{-1} scan rate, and Ragone plot (I).

In addition, the expanded interlayer space could facilitate the fast reversible redox reaction of NiMn-G-LDH nanosheets. Thus, glucose intercalated NiMn-G-LDH@$NiCo_2S_4$@CFC electrode material-based FSAS could be a novel energy storage device for wearable supercapacitor applications. Subsequently, a combination of transition metal oxide and metal hydroxide has high theoretical capacitance and superior redox reactions. In addition, the architecture of the electrode material as a core-shell structure provides large active sites for electrochemical energy storage. Z. Tan et al. developed the novel core-shell architecture of binary metal hydroxides anchored on copper foam supported on 3D copper oxide (CuO) frameworks [73] and Fig. 3.10 A represents the schematic illustration of the synthesis of binder-free 3D CuO@ZnCo-OH core-shell heterostructure. Initially, the $Cu(OH)_2$ nanorod array was directly grown from the CF substrate by immersing it in NaOH. Consequently, the formed $Cu(OH)_2$ nanorods subjected to a thermal annealing process, result in the formation of CuO.

The nanorod morphology was retained even after the annealing treatment. But the surface of nanorods turns out to be rough and the diameter of the nanorods has decreased. Afterward, the cathodic electrodeposition was applied to develop the CuO@ZnCo-OH core-shell structure. Briefly, the Co^{2+} ions in the electrolyte solution were oxidized into Co^{3+} by hydrogen peroxide. Upon applying the cathodic potential at −1.0 V to the CuO electrode, the reduction of nitrate ions (NO_3^-) and H_2O from the electrolyte solution leads to the formation of OH^- ions. Further, OH^- ions co-precipitated with metal cations such as Zn^{2+}, Co^{2+}, and Co^{3+}, resulting in the formation of CuO@ZnCo-OH core-shell heterostructure. The electrochemical redox reaction involved in the electrodeposition process is given as:

$$2Co^{2+} + H_2O_2 \rightarrow 2Co^{3+} + 2OH^-$$

$$2NO_3^- + H_2O + 2e^- \rightarrow NO_2^- + 2OH^-$$

$$2H_2O + 2e^- \rightarrow H_2 + 2OH^-$$

$$Zn^{2+} + Co^{2+} + Co^{3+} + OH^- \rightarrow ZnCo - OH$$

For comparison, the molar ratio of Zn^{2+}/Co^{2+} has been varied and the samples are denoted as CuO@Zn_1Co_0–OH, CuO@Zn_1Co_1–OH, CuO@Zn_1Co_2–OH, CuO@Zn_1Co_3–OH, and CuO@Zn_0Co_1–OH. Further, TEM images in Fig. 3.10 B and C show a detailed microstructure of ZnCo-OH nanosheets that are uniformly coated on the surface of CuO nanorods. Hence, the core-shell heterostructure is successfully formed. Noticeably, CuO@Zn_1Co_2–OH electrode material exhibits a maximum capacitive performance of 2.634 F cm^{-2} at 5 mA cm^{-2} current density in the three-electrode system. It is primarily due to the synergistic effect between the CuO core and Zn_1Co_2–OH shell leading to the enhancement of electrochemical redox reactions.

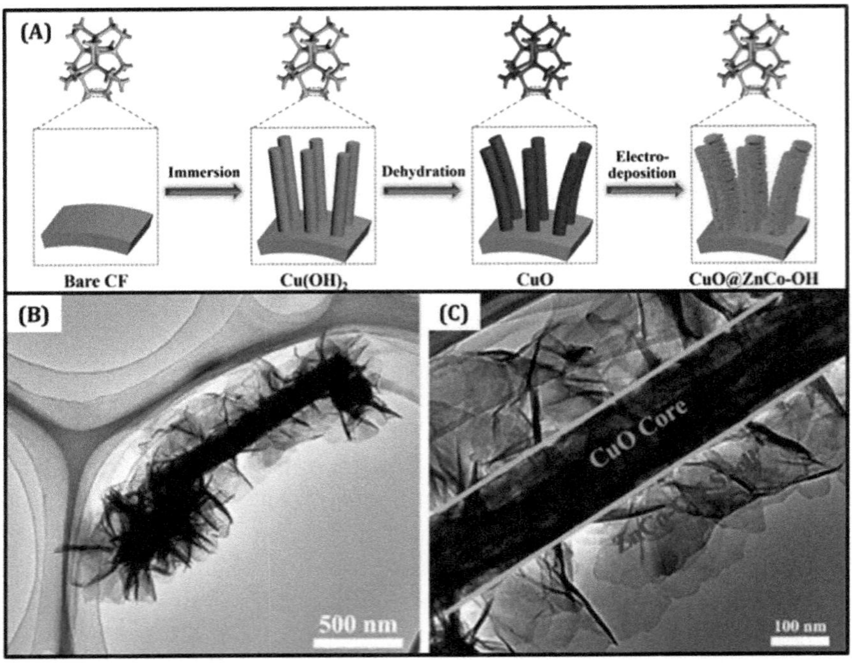

Figure 3.10 Schematic illustration of the synthesis (A) and (B & C) TEM images of different magnifications of the hierarchical CuO@ZnCo–OH core-shell structure.

On the other hand, the potential application of CuO@Zn_1Co_2–OH electrode was checked by assembling an asymmetric supercapacitor, where Zn_1Co_2–OH acts as the positive electrode and loofah sponge-derived carbon as the negative electrode. The assembled asymmetric supercapacitor is denoted as CuO@Zn_1Co_2–OH//LSDC. The comparative CV curves of CuO@Zn_1Co_2–OH (0 to 0.5 V) and LSDC (–1 to 0 V) electrode at 5 mV s^{-1} scan rate in the three-electrode system was shown in Fig. 3.11 A. Further, the operating potential window was extended to 1.5 V.

Moreover, the $CuO@Zn_1Co_2$-OH//LSDC device has delivered a pair of redox peaks in both CV and GCD profiles at different potential windows (Fig. 3.11 B and C). Similarly, the CV curves obtained in the potential range between 0 and 1.5 V also exhibit a well-resolved redox peak up to a higher scan rate at 50 mV s^{-1}, indicating the pseudocapacitive nature of the $CuO@Zn_1Co_2$-OH// LSDC device.

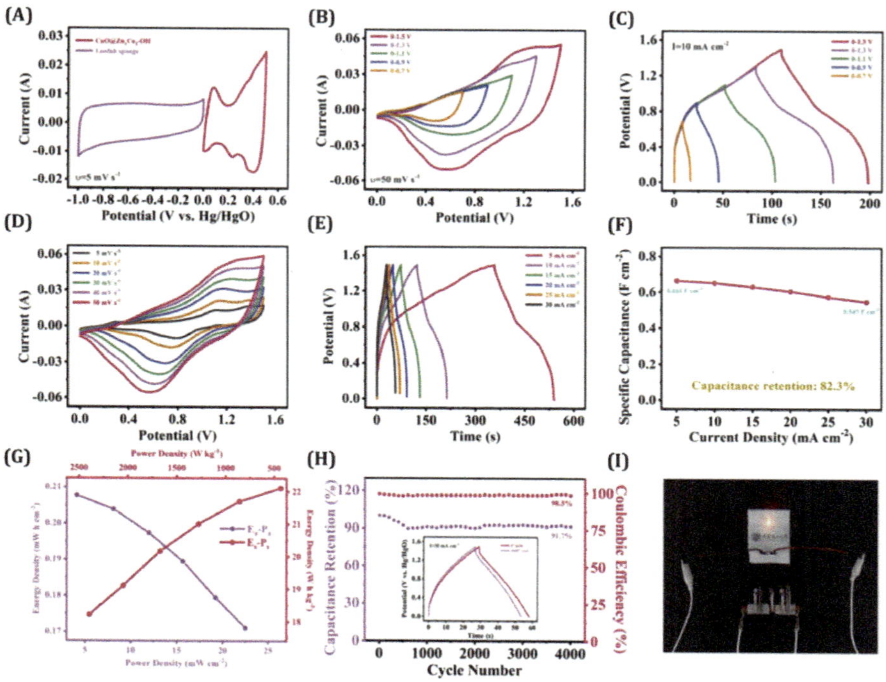

Figure 3.11 Electrochemical performance of the $CuO@Zn_1Co_2$–OH//LSDC device. Comparison of CV curves for $CuO@Zn_1Co_2$–OH and LSDC electrodes at a scan rate of 5 mV s^{-1} in a three-electrode system (A), CV curves of the device in different voltage windows at a scan rate of 50 mV s^{-1} (B), GCD curves of the device in different voltage windows under a current density of 10 mA cm^{-2} (C), CV curves of the device collected at different scan rates ranging from 5 to 50 mV s^{-1} (D), GCD curves of the device collected under different current densities ranging from 5 to 30 mA cm^{-2} (E), areal capacitances versus different current densities ranging from 5 to 30 mA cm^{-2} for the device (F), Ragone plot (G), cycling stability of the device under a constant current density of 30 mA cm^{-2} (inset shows the GCD curves of the first and the last GCD cycles) (H), and optical images of a red LED lit by two charged solid-state asymmetric supercapacitors in series (I).

In addition, the GCD curve at 5 mA cm^{-2} current density exhibits a large discharge curve with a high capacitance of 0.664 F cm^{-2} at 5 mA cm^{-2} (Fig. 3.11 D and E) and retains the specific capacitance of 0.547 F cm^{-2} at 30 mA cm^{-2} with the capacitance retention of 82.3%. It is seen in Fig. 3.11 F, where the fabricated device shows a high energy density of 0.21 mWh cm^{-2} at a power density of 4.08 mW cm^{-2} (Fig. 3.11 G). Also maintained the energy density of 0.17 mWh cm^{-2} when the power density reached 22.49 mW cm^{-2} and explored the cycling stability study at 30 mA cm^{-2} for 4000 cycles (Fig. 3.11 H). Particularly, the $CuO@Zn_1Co_2$–OH//LSDC device has maintained 91.7% of the initial capacitance even after 4000 GCD cycles. Figure 3.11 H shows the GCD curves of the first and last cycles, which inferred the excellent lifespan of the $CuO@Zn_1Co_2$–OH//LSDC device. Besides, the device can lighten up the 2.5 V LED bulb at 20 mA applied current density. Thus, $CuO@Zn_1Co_2$–OH electrode material offers a new way to produce advanced electrochemical energy storage devices by tuning the heterostructure of the core-shell material.

3.5 Summary and Future Aspects

LDH-based 2D nanomaterials are promising electrode materials, due to their facile preparation, high crystallinity, good tunability, and cost-effectiveness. Therefore, designing and synthesizing highly active and low-cost LDH-based electrode materials are essential to improve the energy storage capacity of the supercapacitor and batteries. In this review, we summarized recent developments of LDH-based electrochemical energy storage devices. Notably, miniaturized energy storage devices have received more attention, due to their flexibility, size, stabilization, and easy integration. Moreover, the understanding of the synergistic effect between two or more LDH components for effective mass diffusion and electron transfer reactions is very challenging and significant. It is also needed to probe the size and shape of the energy storage devices. In addition, the development of ingenious preparation methods of 2D or 3D LDH materials deliberately can illustrate an enhanced electrochemical storage property. Further, tuning the electronic structure of LDH materials could offer detailed insights into the electrochemical reaction process. Future efforts are to be directed at making 2D or 3D LDH materials obtained by rational design and ingenious processes for microelectronic energy storage devices.

References

1. N.K. Sharma, N. Tyagi, V. Rana, Renewable energy scenario in India: A current status, *Int. J. Appl. Eng. Res.* 14 (2019) 150–154.
2. A. Khaligh, O.C. Onar, *Energy Sources*, 4th ed., Elsevier Inc., 2018. doi:10.1007/978-3-030-02822-0_2.
3. I. Khan, K. Saeed, I. Khan, Nanoparticles: Properties, applications and toxicities, *Arab. J. Chem.* 12 (2017) 908–931. doi:10.1016/j.arabjc.2017.05.011.
4. K. Jiang, Q. Weng, Miniaturized energy storage devices based on two-dimensional materials, *ChemSusChem.* 13 (2020) 1420–1446. doi:10.1002/cssc.201902520.
5. Z. Gholamvand, D. Mcateer, A. Harvey, C. Backes, J.N. Coleman, Electrochemical applications of 2-dimensional nanosheets: The effect of nanosheet length and thickness, *Chem. Mater.* 28 (2016) 2641–2651. doi:10.1021/acs.chemmater.6b00009.
6. S. Hu, S.R. Das, H. Monshat, 2D materials: Applications for electrochemical energy storage devices, G. Balasubramanian (Ed.), *Adv. Nanomater.* (2018) 61–75. doi:10.1007/978-3-319-64717-3.
7. Z. Wu, L. Li, J. Yan, X. Zhang, Materials design and system construction for conventional and new-concept supercapacitors, *Adv. Sci.* 4 (2017) 1600382.
8. G. Aruchamy, S. Thangavelu, Bifunctional $CoSn(OH)_6/MnO_2$ composite for solid-state asymmetric high power density supercapacitor and for an enhanced OER, *Electrochim. Acta.* 344 (2020) 136141. doi:10.1016/j.electacta.2020.136141.
9. Z.S. Iro, C. Subramani, S.S. Dash, A brief review on electrode materials for supercapacitor, *Int. J. Electrochem. Sci.* 11 (2016) 10628–10643.
10. E.E. Miller, Y. Hua, F.H. Tezel, Materials for energy storage: Review of electrode materials and methods of increasing capacitance for supercapacitors, *J. Energy Storage.* 20 (2018) 30–40. doi:10.1016/j.est.2018.08.009.

11. Y. Wang, Y. Song, Y. Xia, Electrochemical capacitors: Mechanism, materials, systems, characterization and applications, *Chem. Soc. Rev.* 45 (2016) 5925–5950. doi:10.1039/C5CS00580A.

12. R. Wang, P. Jia, Y. Yang, N. An, Y. Zhang, H. Wu, Z. Hu, Ruthenium oxide/reduced graphene oxide nanoribbon composite and its excellent rate capability in supercapacitor application, *Chinese J. Chem.* 34 (2016) 114–122. doi:10.1002/cjoc.201500595.

13. V.H. Nguyen, J. Shim, Ultrasmall SnO_2 nanoparticle-intercalated graphene @ polyaniline composites as an active electrode material for supercapacitors in different electrolytes, *Synth. Met.* 207 (2015) 110–115. doi:10.1016/j.synthmet.2015.06.010.

14. P. Hao, Z. Zhao, J. Tian, H. Li, Y. Sang, G. Yu, H. Cai, H. Liu, C.P. Wong, A. Umar, Hierarchical porous carbon aerogel derived from bagasse for high performance supercapacitor, *Nanoscale.* 6 (2014) 12120–12129. doi:10.1039/C4NR03574G.

15. I. Chakraborty, N. Chakrabarty, A. Senapati, A.K. Chakraborty, CuO@NiO/Polyaniline/MWCNT nanocomposite as high-performance electrode for supercapacitor, *J. Phys. Chem. C.* 122 (2018) 27180–27190. doi:10.1021/acs.jpcc.8b08091.

16. Q. Wang, Y. Zhang, H. Jiang, T. Hu, C. Meng, In-situ generated $Ni_3Si_2O_5(OH)_4$ on mesoporous heteroatom-enriched carbon derived from natural bamboo leaves for high performance supercapacitors, *ACS Appl. Energy Mater.* 1 (2018) 2296–3409. doi:10.1021/acsaem.8b00556.

17. A. Sarkar, A.K. Chakraborty, S. Bera, S. Krishnamurthy, Novel hydrothermal synthesis of CoS_2/MWCNT nanohybrid electrode for supercapacitor: A systematic investigation on the influence of MWCNT, *J. Phys. Chem. C.* 122 (2018) 18237–18246. doi:10.1021/acs.jpcc.8b04137.

18. J.J. William, I.M. Babu, G. Muralidharan, Spongy structured α-$Ni(OH)_2$: Facile and rapid synthesis for supercapattery applications, *Mater. Lett.* 238 (2018) 35–37. doi:10.1016/j.matlet.2018.11.136.

19. J.N. Tiwari, R.N. Tiwari, K.S. Kim, Zero-dimensional, one-dimensional, two-dimensional and three-dimensional nanostructured materials for advanced electrochemical energy devices, *Prog. Mater. Sci.* 57 (2012) 724–803. doi:10.1016/j.pmatsci.2011.08.003.

20. X. Li, J. Wang, One-dimensional and two-dimensional synergized nanostructures for high-performing energy storage and conversion, *InfoMat.* 2 (2020) 3–32. doi:10.1002/inf2.12040.

21. D. Jiang, X. Li, Q. Jia, Design of two-dimensional layered double hydroxide nanosheets embedded with Fe_3O_4 for highly selective enrichment and isotope labeling of phosphopeptides, *ACS Sustain. Chem. Eng.* 7 (2019) 421–429. doi:10.1021/acssuschemeng.8b03806.

22. F. Giannazzo, G. Greco, F. Roccaforte, S.S. Sonde, Vertical transistors based on 2D materials: Status and prospects, *Crystals.* 8 (2018) 70.

23. S. Zheng, X. Shi, P. Das, Z. Wu, X. Bao, The road towards planar microbatteries and micro-supercapacitors: From 2D to 3D device geometries, *Adv. Mater.* 31 (2019) 1900583. doi:10.1002/adma.201900583.

24. X. Shi, Z. Shuai, W. Xinhe, Recent advancements and perspective of high-performance printed power sources with multiple form factors, *Electrochem. Energy Rev.* 3 (2020) 581–612. doi:10.1007/s41918-020-00071-6.

25. Y.-Z. Zhang, Y. Wang, T. Cheng, L.-Q. Yao, X. Li, W.-Y. Lai, W. Huang, Printed supercapacitors: Materials, printing and applications, *Chem. Soc. Rev.* 48 (2019) 3229. doi:10.1039/c7cs00819h.

26. M.A. Iqbal, L. Sun, A.T. Barrett, M. Fedel, Layered double hydroxide protective films developed on aluminum and aluminum alloys: Synthetic methods and anti-corrosion mechanisms, *Coatings.* 10 (2020) 428.

27. R.M.M. Santos, J. Tronto, V. Briois, C. V Santilli, Thermal decomposition and recovery properties of ZnAl–CO_3 layered double hydroxide for anionic dye adsorption: Insight into the aggregative nucleation and growth mechanism of the LDH memory effect, *J. Mater. Chem. A.* 5 (2017) 9998–10009. doi:10.1039/C7TA00834A.

28. G. Mishra, B. Dash, S. Pandey, Layered double hydroxides: A brief review from fundamentals to application as evolving biomaterials, *Appl. Clay Sci.* 153 (2018) 172–186. doi:10.1016/j.clay.2017.12.021.

29. T. Hur, T.X. Phuoc, M.K. Chyu, New approach to the synthesis of layered double hydroxides and associated ultrathin nanosheets in de-ionized water by laser ablation, *J. Appl. Phys.* 108 (2010) 114312. doi:10.1063/1.3518510.

30. H.R. Prakruthi, B.M. Chandrashekara, B.S.J. Prakash, Y.S. Bhat, Microwave rehydrated Mg-Al-LDH as base catalyst for the acetalization of glycerol, *Catal. Sci. Technol.* 5 (2015) 3667–3674. doi:10.1039/C5CY00347D.

31. F. Wang, S. Sun, Y. Xu, T. Wang, R. Yu, H. Li, High performance asymmetric supercapacitor based on cobalt nickle iron-layered

double hydroxide/carbon nanofibres and activated carbon, *Sci. Rep.* 7 (2017) 1–11.

32. F. Prinetto, G. Ghiotti, P. Gra, D. Tichit, Synthesis and characterization of sol-gel Mg/Al and Ni/Al layered double hydroxides and comparison with co-precipitated samples, *Microporus Mesoporus Mater.* 39 (2000) 229–247.
33. K. Inomata, M. Ogawa, Preparation and properties of Mg/Al layered double hydroxide–oleate and –stearate intercalation compounds, *Bull. Chem. Soc. Jpn.* 79 (2006) 336–342. doi:10.1246/cl.2005.810.
34. A. Malak-polaczyk, C. Vix-guterl, E. Frackowiak, Carbon/layered double hydroxide (LDH) composites for supercapacitor application, *Energy and Fuels.* 24 (2010) 3346–3351. doi:10.1021/ef901505c.
35. G. V Manohara, Exfoliation of layered double hydroxides (LDHs): A new route to mineralize atmospheric CO_2, *RSC Adv.* 4 (2014) 46126–46132. doi:10.1039/C4RA08865D.
36. J. Sun, Z. Li, J. Wang, W. Hong, P. Gong, P. Wen, Z. Wang, S. Yang, Ni/Bi battery based on $Ni(OH)_2$ nanoparticles/graphene sheets and Bi_2O_3 rods/ graphene sheets with high performance, *J. Alloys Compd.* 643 (2015) 231–238. doi:10.1016/j.jallcom.2015.04.137.
37. F. Wang, X. Wang, Z. Chang, Y. Zhu, L. Fu, X. Liu, Y. Wu, Electrode materials with tailored facets for electrochemical energy storage, *Nanoscale Horizons.* 1 (2016) 272–289. doi:10.1039/C5NH00116A.
38. L. He, T. Hong, X. Hong, X. Liao, Y. Chen, W. Zhang, H. Liu, W. Luo, L. Mai, Ultrastable high-energy on-chip nickel–bismuth microbattery powered by crystalline Bi anode and Ni–Co hydroxide cathode, *Energy Technol.* 7 (2019) 1900144. doi:10.1002/ente.201900144.
39. H. Ma, S. Du, H. Tao, T. Li, Y. Zhang, Three-dimensionally integrated carbon tubes/MoS_2 with reduced graphene oxide foam as a binder-free anode for lithium ion battery, *J. Electroanal. Chem.* 823 (2018) 307–314. doi:10.1016/j.jelechem.2018.06.033.
40. Z. Liu, X. Tian, X. Xu, L. He, M. Yan, C. Han, Y. Li, W. Yang, L. Mai, Capacitance and voltage matching between MnO_2 nanoflake cathode and Fe_2O_3 nanoparticle anode for high-performance asymmetric micro-supercapacitors, *Nano Res.* 10 (2017) 2471–2481. doi:10.1007/s12274-017-1451-4.
41. L. Zhang, D. Dearmond, N.T. Alvarez, R. Malik, N. Oslin, C. Mcconnell, P.K. Adusei, Y. Hsieh, V. Shanov, Flexible micro-supercapacitor based on graphene with 3D structure, *Small.* (2017) 1603114. doi:10.1002/smll.201603114.

42. W. Liu, C. Lu, X. Wang, R.Y. Tay, K. Tay, High-performance micro-supercapacitors based on two-dimensional graphene/manganese dioxide/silver nanowire ternary hybrid film, *ACS Nano.* 9 (2015) 1528–1542. doi:10.1021/nn5060442.

43. S. Palchoudhury, K. Ramasamy, R.K. Gupta, A. Gupta, Flexible supercapacitors: A materials perspective, *Front. Mater.* 5 (2019) 1–9. doi:10.3389/fmats.2018.00083.

44. Y. Wang, X. Hong, Y. Guo, Y. Zhao, X. Liao, X. Liu, Q. Li, L. He, L. Mai, Wearable textile-based Co–Zn alkaline microbattery with high energy density and excellent reliability, *Small.* (2020) 2000293. doi:10.1002/smll.202000293.

45. X. Pu, M. Liu, L. Li, S. Han, X. Li, C. Jiang, C. Du, J. Luo, W. Hu, Z.L. Wang, Wearable textile-based in-plane microsupercapacitors, *Adv. Energy Mater.* (2016) 1601254. doi:10.1002/aenm.201601254.

46. X. Pu, L. Li, M. Liu, C. Jiang, C. Du, Z. Zhao, W. Hu, Z.L. Wang, Wearable self-charging power textile based on flexible yarn supercapacitors and fabric nanogenerators, *Adv. Mater.* 28 (2016) 98–105. doi:10.1002/adma.201504403.

47. G. Sun, H. Yang, G. Zhang, J. Gao, X. Jin, Y. Zhao, L. Jiang, L. Qu, A capacity recoverable zinc-ion micro-supercapacitor, *Energy Environ. Sci.* 11 (2018) 3367–3374. doi:10.1039/c8ee02567c.

48. Y. Zeng, Y. Meng, Z. Lai, X. Zhang, M. Yu, P. Fang, M. Wu, Y. Tong, X. Lu, An ultrastable and high-performance flexible fiber-shaped Ni–Zn battery based on a Ni–NiO heterostructured nanosheet cathode, *Adv. Mater.* (2017) 1702698. doi:10.1002/adma.201702698.

49. Z. Wang, J. Cheng, Q. Guan, H. Huang, Y. Li, J. Zhou, W. Ni, B. Wang, S. He, H. Peng, All-in-one fiber for stretchable fiber-shaped tandem supercapacitors, *Nano Energy.* 45 (2017) 210–219. doi:10.1016/j.nanoen.2017.12.054.

50. Z. Lv, Y. Luo, Y. Tang, J. Wei, Z. Zhu, X. Zhou, W. Li, Y. Zeng, W. Zhang, Y. Zhang, D. Qi, S. Pan, X.J. Loh, X. Chen, Editable supercapacitors with customizable stretchability based on mechanically strengthened ultralong MnO_2 nanowire composite, *Adv. Mater.* (2017) 1704531. doi:10.1002/adma.201704531.

51. P. Zhang, Y. Li, G. Wang, F. Wang, S. Yang, F. Zhu, X. Zhuang, O.G. Schmidt, X. Feng, Zn-ion hybrid micro-supercapacitors with ultrahigh areal energy density and long-term durability, *Adv. Mater.* (2019) 1806005. doi:10.1002/adma.201806005.

52. J. Yun, Y. Lim, H. Lee, G. Lee, H. Park, S.Y. Hong, S.W. Jin, Y.H. Lee, S.-S. Lee, J.S. Ha, A patterned graphene/ZnO UV sensor driven by

integrated asymmetric micro-supercapacitors on a liquid metal patterned foldable paper, *Adv. Funct. Mater.* (2017) 1700135. doi:10.1002/adfm.201700135.

53. J.Q. Xie, Y.Q. Ji, J.H. Kang, J.L. Sheng, D.S. Mao, X.Z. Fu, R. Sun, C.P. Wong, In situ growth of $Cu(OH)_2$@FeOOH nanotube arrays on catalytically deposited Cu current collector patterns for high-performance flexible in-plane micro-sized energy storage devices, *Energy Environ. Sci.* 12 (2019) 194–205. doi:10.1039/c8ee01979g.
54. X. Shi, S. Pei, F. Zhou, W. Ren, H.M. Cheng, Z.S. Wu, X. Bao, Ultrahigh-voltage integrated micro-supercapacitors with designable shapes and superior flexibility, *Energy Environ. Sci.* 12 (2019) 1534–1541. doi:10.1039/c8ee02924e.
55. M. Li, J. Meng, Q. Li, M. Huang, X. Liu, K.A. Owusu, Z. Liu, L. Mai, Finely crafted 3D electrodes for dendrite-free and high-performance flexible fiber-shaped Zn–Co batteries, *Adv. Funct. Mater.* 28 (2018) 1–10. doi:10.1002/adfm.201802016.
56. J. Liu, C. Guan, C. Zhou, Z. Fan, Q. Ke, G. Zhang, C. Liu, J. Wang, A flexible quasi-solid-state nickel–zinc battery with high energy and power densities based on 3D electrode design, *Adv. Mater.* 28 (2016) 8732–8739. doi:10.1002/adma.201603038.
57. C. Xu, J. Liao, C. Yang, R. Wang, D. Wu, P. Zou, Z. Lin, B. Li, F. Kang, C.P. Wong, An ultrafast, high capacity and superior longevity Ni/Zn battery constructed on nickel nanowire array film, *Nano Energy.* 30 (2016) 900–908. doi:10.1016/j.nanoen.2016.07.035.
58. M.F. El-Kady, V. Strong, S. Dubin, R.B. Kaner, Laser scribing of high-performance and flexible graphene-based electrochemical capacitors, *Science.* 335 (2012) 1326–1330. doi:10.1126/science.1216744.
59. B. Anothumakkool, R. Soni, S.N. Bhange, S. Kurungot, Novel scalable synthesis of highly conducting and robust PEDOT paper for a high performance flexible solid supercapacitor, *Energy Environ. Sci.* 8 (2015) 1339–1347. doi:10.1039/c5ee00142k.
60. C. Guan, W. Zhao, Y. Hu, Q. Ke, X. Li, H. Zhang, J. Wang, High-performance flexible solid-state Ni/Fe battery consisting of metal oxides coated carbon cloth/carbon nanofiber electrodes, *Adv. Energy Mater.* 6 (2016) 1–9. doi:10.1002/aenm.201601034.
61. K.A. Owusu, L. Qu, J. Li, Z. Wang, K. Zhao, C. Yang, K.M. Hercule, C. Lin, C. Shi, Q. Wei, L. Zhou, L. Mai, Low-crystalline iron oxide hydroxide nanoparticle anode for high-performance supercapacitors, *Nat. Commun.* 8 (2017) 14264. doi:10.1038/ncomms14264.

62. F. Pei, L. Lin, D. Ou, Z. Zheng, S. Mo, X. Fang, N. Zheng, Self-supporting sulfur cathodes enabled by two-dimensional carbon yolk-shell nanosheets for high-energy-density lithium-sulfur batteries, *Nat. Commun.* 8 (2017) 1–10. doi:10.1038/s41467-017-00575-8.

63. S. Chen, X. Han, J. Luo, J. Liao, J. Wang, Q. Deng, Z. Zeng, S. Deng, In situ transformation of LDH into hollow cobalt-embedded and N-doped carbonaceous microflowers as polysulfide mediator for lithium-sulfur batteries, *Chem. Eng. J.* 385 (2020) 123457. doi:10.1016/j.cej.2019.123457.

64. Y.J. Li, J.M. Fan, M. Sen Zheng, Q.F. Dong, A novel synergistic composite with multi-functional effects for high-performance Li-S batteries, *Energy Environ. Sci.* 9 (2016) 1998–2004. doi:10.1039/c6ee00104a.

65. S. Liu, J. Li, X. Yan, Q. Su, Y. Lu, J. Qiu, Z. Wang, X. Lin, J. Huang, R. Liu, B. Zheng, L. Chen, R. Fu, D. Wu, Superhierarchical cobalt-embedded nitrogen-doped porous carbon nanosheets as two-in-one hosts for high-performance lithium–sulfur batteries, *Adv. Mater.* 30 (2018) 1–9. doi:10.1002/adma.201706895.

66. S. Peng, L. Li, C. Li, H. Tan, R. Cai, H. Yu, S. Mhaisalkar, M. Srinivasan, S. Ramakrishna, Q. Yan, In situ growth of $NiCo_2S_4$ nanosheets on graphene for high-performance supercapacitors, *Chem. Commun.* 49 (2013) 10178–10180. doi:10.1039/c3cc46034g.

67. S. Chen, C. Lu, L. Liu, M. Xu, J. Wang, Q. Deng, Z. Zeng, S. Deng, A hierarchical glucose-intercalated NiMn-G-LDH@$NiCo_2S_4$ core-shell structure as a binder-free electrode for flexible all-solid-state asymmetric supercapacitors, *Nanoscale.* 12 (2020) 1852–1863. doi:10.1039/c9nr09083e.

68. H. Liang, J. Lin, H. Jia, S. Chen, J. Qi, J. Cao, T. Lin, W. Fei, J. Feng, Hierarchical NiCo-LDH/NiCoP@NiMn-LDH hybrid electrodes on carbon cloth for excellent supercapacitors, *J. Mater. Chem. A.* 6 (2018) 15040–15046. doi:10.1039/c8ta05065a.

69. M. Liang, M. Zhaoa, H. Wanga, J. Shena, X. Song, Enhanced cycling stability of hierarchical $NiCo_2S_4$@$Ni(OH)_2$@PPy core-cell nanotube arrays for aqueous asymmetric supercapacitors, *J. Mater. Chem. A.* 6 (2018) 2482–2493. doi:10.1039/C7TA10413H.

70. Y. Wen, S. Peng, Z. Wang, J. Hao, T. Qin, S. Lu, J. Zhang, D. He, X. Fan, G. Cao, Facile synthesis of ultrathin $NiCo_2S_4$ nano-petals inspired by blooming buds for high-performance supercapacitors yuxiang, *J. Mater. Chem. A.* 5 (2017) 7144–7152. doi:10.1039/C7TA01326D.

71. X. Li, H. Wu, C. Guan, A.M. Elshahawy, Y. Dong, S.J. Pennycook, J. Wang, (Ni,Co)Se_2/NiCo-LDH core/shell structural electrode with

the cactus-like (Ni,Co)Se_2 core for asymmetric supercapacitors, *Small.* 15 (2019) 1803895. doi:10.1002/smll.201803895.

72. X. Guan, M. Huang, L. Yang, G. Wang, X. Guan, Facial design and synthesis of CoS_x/Ni-Co LDH nanocages with rhombic dodecahedral structure for high-performance asymmetric supercapacitors, *Chem. Eng. J.* 372 (2019) 151–162. doi:10.1016/j.cej.2019.04.145.

73. G. Liu, X. Song, S. Zhang, X. Chen, S. Liu, Y. Meng, Z. Tan, Hierarchical CuO@ZnCo–OH core-shell heterostructure on copper foam as three-dimensional binder-free electrodes for high performance asymmetric supercapacitors, *J. Power Sources.* 465 (2020) 228239. doi:10.1016/j.jpowsour.2020.228239.

Chapter 4

Graphitic Carbon Nitride for Energy Storage Applications

Sreejith P. Babu,[a] Himadri Tanaya Das,[b] Nibagani Naresh,[c] Kamatam Hari Prasad,[d] and Nigamananda Das[e]

[a]*Electrochemistry for Energy & Environment Group, School of Physical and Chemical Sciences, North-West University, Potchefstroom 2520, South Africa*
[b]*Department of Chemical Engineering, National Taiwan University, Taipei 10617, Taiwan*
[c]*Department of Physics, Pondicherry University, Pondicherry 605014, India*
[d]*Department of Physics, Institute of Aeronautical Engineering, Hyderabad 500043, Telangana India*
[e]*Department of Chemistry, Utkal University, Vanivihar, Bhubaneswar 751004, Odisha, India*

kuttan421@gmail.com, himadritanaya1991@gmail.com

With constant improvement in electrochemical energy storage systems, accumulating energy harnessed from renewable sources has become more effective, affordable, and accessible. Commercialized electrochemical devices such as batteries and supercapacitors have rapidly taken over as energy solutions for medical devices and consumer electronics. To extend the existing electrochemical energy storage infrastructure for permeation into automotive and large-scale applications, as a measure of mitigating carbon emissions, significant steps need to be adopted. The development and inclusion of high-capacity energy storage

Layered Materials: Synthesis and Applications in Energy and Environmental Remediation
Edited by Alagarsamy Pandikumar, Kandasamy Jothivenkatachalam, and Subramanian Moscow

ISBN 978-981-5129-17-5 (Hardcover), 978-1-003-50831-1 (eBook)
www.jennystanford.com

solutions have been studied using materials based on soil-rich elements as potential electrodes. In this context, the emergence of graphitic carbon nitride (g-C_3N_4) plays a significant role in improving the energy storage performance of batteries and supercapacitors. The chapter presents an extensive and insightful discussion on the adoption of g-C_3N_4-based materials and the improvements brought about. This chapter addresses the general merits and demerits of the g-C_3N_4 for electrochemical energy storage applications. Improvements in the suitability and versatility of the g-C_3N_4 for commercial usage in Li-ion, Na-ion, metal-air, and Li–S batteries and supercapacitors have been suggested.

4.1 Introduction

To reduce dependence and mitigate the adverse environmental effects brought in by fossil fuel resources, the use of renewable energy technologies for efficient energy storage is receiving considerable emphasis [1–4]. In addition to that, establishing energy storage technologies that do not compromise the environment is key to integrating renewable energy technologies with the existing grid. The reliability, portability, and adaptability of electrochemical energy storage systems such as batteries and supercapacitors render them quite attractive for the purpose. The application of such energy storage devices has already become significant in daily life, spanning the realms of consumer electronics to transportation. However, challenges related to the fabrication, scale-up, restricted fast charging, capacity limitations, and limited cycle life of the energy storage devices require extensive research to improve upon.

Carbon nanomaterials, ranging from graphene to carbon nanodots, are interesting stimulants of research, particularly in the fields of environmental management, energy transfer, and energy storage. Predominantly, the layered structure of materials such as graphene, phosphorene, MXenes, graphitic carbon nitrides, and borides offer a good platform for fast electrochemical reactions in various energy storage applications [5–7]. Among the layered materials, g-C_3N_4 polymers with a general composition of $(C_3N_3H)_n$ have been significant in the electrochemical energy storage literature [8]. The g-C_3N_4 is composed of stacked 2D planar sheets with an interlayer gap of 0.319 nm. The layered graphitic structure of the g-C_3N_4 sheets arises due to the π-conjugated electronic structures comprising sp^2 hybridized atoms of carbon and nitrogen [9]. The ring-patterned tri-s-triazine structure of g-C_3N_4 is responsible for the high thermal and chemical stability (600 °C in acidic and alkaline environments), rendering it a stable allotrope [10, 11]. The higher packaging density of g-C_3N_4 also leads to closely connected layers and enhances charge transport in the direction perpendicular to the planes, contrary to that in graphite. To add to it, the nanosheet morphology of g-C_3N_4 adds to the electrochemical surface area and thereby large energy storage capacity. Owing to its structural and chemical stability, g-C_3N_4-based electrodes support higher diffusion of ions into layers, whilst undergoing minimum strain. Following the high electrochemical performance of g-C_3N_4, it can be adopted for numerous energy storage applications. In comparison to other carbon-based compounds, the synthesis of g-C_3N_4 is also relatively simple. The g-C_3N_4 can be synthesized by thermal polymerization of nitrogen-rich compounds or by-products of pre-bonded C-N core structures including urea, dicyandiamide, cyanamide, thiourea, guanidinium chloride, thiocyanide guanidine, and thiourea oxide [12–15]. Alteration of precursors and experimental parameters have been used to achieve desired structures such as 3D bulk, 2D foil/nanosheet, 1D nanotubes, nanowires or nanorods, and 0D quantum dots [16–18].

Nevertheless, high contact resistance, bandgap, and limited bulk g-C_3N_4 surface area results in low conductivity and catalytic activity, thus affecting the efficiency of electrochemical energy storage [19]. Energy storage applications comprising g-C_3N_4 nanocomposites with promising fast charge/discharge, healthy operation, long lifecycles, and high-power density are being researched. In various energy storage devices such as lithium-ion batteries (LIBs), sodium-ion batteries (NIBs), metal–air batteries (MABs), Li–O_2 batteries, zinc–air batteries (ZABs), lithium–sulfur (Li–S) batteries, and supercapacitors (SCs), the integration of g-C_3N_4 to form nanocomposites with different carbon compounds, metal oxides/hydroxides, or metal chalcogenides has been utilized [10, 12, 14]. In this chapter, g-C_3N_4-based electrode systems that have been used in all of the above energy storage applications have been discussed in detail.

4.2 Graphitic Carbon Nitride g-C_3N_4-Based Electrodes for LIBs

The demand for LIBs has increased following the requirement for safe, lightweight, pollution-free, and high-energy-density alternatives to fossil fuels [20]. Commercial LIBs typically use a graphite anode and an intercalated Li compound such as $LiCoO_2$ as a cathode material. Typically, an ideal anode should have the following essential characteristics such as Li adsorption energy above 1.6 eV (bulk Li) and below 3 eV to minimize clustering and allow for Li^+ desorption. High-specific capacity, low-volume change, and retaining structural stability during the Li-ion intercalation and de-intercalation processes, high surface area, and favorable electronic conductivity are also required of an ideal anode material. Nevertheless, in high-power demanding systems such as electric vehicles, the low-energy density of graphite (373 mAh g^{-1}) and $LiCoO_2$ (172 mAh g^{-1}) limits the application of LIBs [21]. Furthermore, the electrochemical capacity of graphite to host the Li^+ ions decreases after prolonged cycling, due to structural decomposition and formation of the solid electrolyte interface (SEI), which barres the movement of Li^+ ions [22].

Recently, there has been much-deliberated interest in 2D materials based on g-C_3N_4. The g-C_3N_4 has garnered attention, owing to its low cost, high surface area, metal-free characteristics, and simple synthesis. Nonetheless, there is also extensive research undergoing to reduce capacity loss and improve conductivity by doping suitable elements in pristine g-C_3N_4. The developments of hybrid binary and ternary nanocomposites with metal oxides, metal sulfides, and different carbon composites are also underway. In this section, discussion on different composites based on g-C_3N_4 as anodes in LIBs and the research being done have been attempted.

4.2.1 Nitrogen-Deficient g-C_3N_4-Based Anodes for LIBs

Since graphite and g-C_3N_4 have almost similar layered structures, the use of g-C_3N_4 as an anode in LIBs was assumed to be quite beneficial, owing to its structural, chemical, and thermal stability, compared to graphite. Pan made a commendable attempt to convey the electrochemical performance of g-C_3N_4 in LIBs by first-principle calculations [23]. He predicted that Li-ion would react with g-C_3N_4 to form $Li_4C_3N_4$ with a theoretical capacity of 1165.3 mAh g^{-1}. However, experimental investigations by Veith et al. on the low reversible capacity for g-C_3N_4 have been largely ascribed to the presence of graphitic C_3N species [24]. During Li-ion insertion and de-insertion in g-C_3N_4 anodes, a substantial decrease in capacity loss is observed, due to the strong binding between g-C_3N species and Li. The authors also observe that to enhance the electronic conductivity of the lithiated g-C_3N_4, an increase in the amount of sp^2 hybridized carbon species is essential. Theoretical calculations from Hankel et al. showed that the stability of the g-C_3N_4 is adversely affected by graphitic nitrogen; whereas, the presence of pyridinic nitrogen is useful for Li-ion storage [25]. In addition to that, the g-C_3N_4 is a poorly conducting semiconductor material, which significantly adds to the irreversible power loss. So, there are two possible ways to improve Li-ion storage in g-C_3N_4, by increasing the amount of pyridinic sites in the g-C_3N_4 and introducing more of the graphene backbone to enhance Li-ion storage capacity and cycling stability. Chen and collaborators reported a magnesiothermic denitriding to synthesize nitrogen-deficient g-C_3N_4 [26]. The process involved combusting g-C_3N_4 and magnesium powder in an Ar atmosphere at 750 °C and reducing the nitrogen content through magnesio-thermic reactions. The surface area of the as-synthesized g-C_3N_4 (579 m^2 g^{-1}) was observed to be 10 times higher than that of g-C_3N_4. After almost 300 cycles, the Li-ion storage capacity of 2753 mAh g^{-1} was obtained for the nitrogen-deficient g-C_3N_4. The improved efficiency and cycling stability of the electrochemical systems are mainly due to reduced nitrogen content within g-C_3N_4, leading to a decrease in CN sites, resulting in a high surface area and higher electronic conductivity. Wu and his collaborators also reported on double the capacity of graphite for lithiated g-C_3N_4

functionalities (542 mAh g^{-1}) and suggested that the electrode could be a possible LIB electrode [27].

4.2.2 Carbon-Based g-C_3N_4 Composites for LIBs

Graphite is used as a standard anode in commercial LIBs because of its flat voltage range, low cost, and layered structure, making it suitable for Li-ion insertion and de-insertion. For every six carbon atoms (LiC_6), graphite is able to accommodate a lithium atom. However, the main disadvantages of graphite-based anodes are low theoretical capacity of around 373 mAh g^{-1}, formation of SEI layer, and structural instability. It was commonly understood that doping can create a large number of active sites/defects, which can enhance electron transport and Li-ion diffusion. Following on the success of doped carbon composites, Huang et al. reported that nitrogen-doped porous carbon nanofibers (CF) exhibited a capacity of 943 mAh g^{-1} at 2 A g^{-1} for LIBs [28]. Tao et al. also demonstrated that g-C_3N_4 composite with N- and P-doped hollow carbon fibers exhibit outstanding capacities of 1030 mAh g^{-1} at 1 A g^{-1} after 1000 cycles and 360 mAh g^{-1} at 10 A g^{-1} after 4000 cycles [29]. Fu et al. synthesized a hybrid composite of rGO and 2D g-C_3N_4 [30]. The synergetic effect and improved electronic conductivity of rGO and 2D g-C_3N_4 hybrid nanocomposites imparted excellent electrochemical performance as anodes for LIBs. This brings us to the conclusion that different nanostructured porous carbon-based materials such as carbon fibers, rGO, graphene, and MWCNT composites with g-C_3N_4 can be potential anodes for LIBs. Many desirable properties of g-C_3N_4 such as Li-storage and chemical and thermal stability coupled with the electronic conductivity and structural stability of different carbon materials are beneficial in developing high-performing anodes for LIBs.

4.2.3 Hybrid Metal Oxide/Sulfide g-C_3N_4 Composites for LIBs

In the quest for suitable anodes for LIBs, transition-metal oxides are also impressive with Li-ion storage capacities that are typically double that of graphite. Unfortunately, anodes based on transition-metal oxides are susceptible to huge volumetric changes during charge–discharge, leading to structural degradation. Metal oxides also suffer from low electronic conductivity, which affects the electrochemical performance of the LIBs. Nonetheless, transition-metal oxide-based anodes can be engineered to have a highly porous morphology with large surface area, reduced volume expansion, shortened Li-ion diffusion pathways, improved electronic conductivity, and increased C-rate capability [31]. The use of Fe_2O_3 composite with carbon, graphene, rGO, etc. as an electrode showcased excellent battery performance such as the one reported by Shi et al. on Fe_2O_3/CN-G composites [32]. They evaluated electrochemical performance and observed a remarkable specific capacity of up to 980 mAh g^{-1} after 50 cycles at a current density of 50 mA g^{-1}. The observed excellent performance corresponds to the incorporation of rGO/g-C_3N_4, which could enhance the conductivity and surface area. Joshi et al. reported a spray-coated synthesis of $ZnFe_2O_4$ composite with g-C_3N_4 as an anode for LIBs with reversible capacity as high as 1235 mAh g^{-1} at a current density of 100 mA g^{-1} after 160 cycles [33]. Versaci et al. successfully integrated g-C_3N_4 with SnO_2 composite, which was used as an anode exhibiting a capacity of 1075 mAh g^{-1} at a current density of 156 mA g^{-1} for 100 cycles (as given in Fig. 4.1) [34].

Silicon as an anode in LIBs has also been attracting attention, due to its high theoretical capacity, which is around 4200 mAh g^{-1}. But its practical application in LIBs is hindered, due to severe volume expansion (more than 300%) and poor electronic conductivity. There are many reports on the preparation of nanosized and different silicon morphologies, which could reduce the volume expansion and improve conductivity. Additionally, the conductivity of nanostructured Si can be improved by coupling with porous carbon, graphene sheets, or MWCNT. In this context, it is imperative to discuss research by Wang et al. on the LIB performance of silicon nanoparticles anchored on a rGO/g-C_3N_4 hybrid matrix [35]. They observed a remarkable capacity of 984.3 mAh g^{-1} at 0.5 C after 500 cycles with 99.8% current efficiency, as shown in Fig. 4.2. The observed outstanding performance is due to the synergistic effect of interface bonding between silicon nanoparticles and 3D porous rGO/g-C_3N_4

nanosheets, ensuring structural stability and improved electronic conductivity.

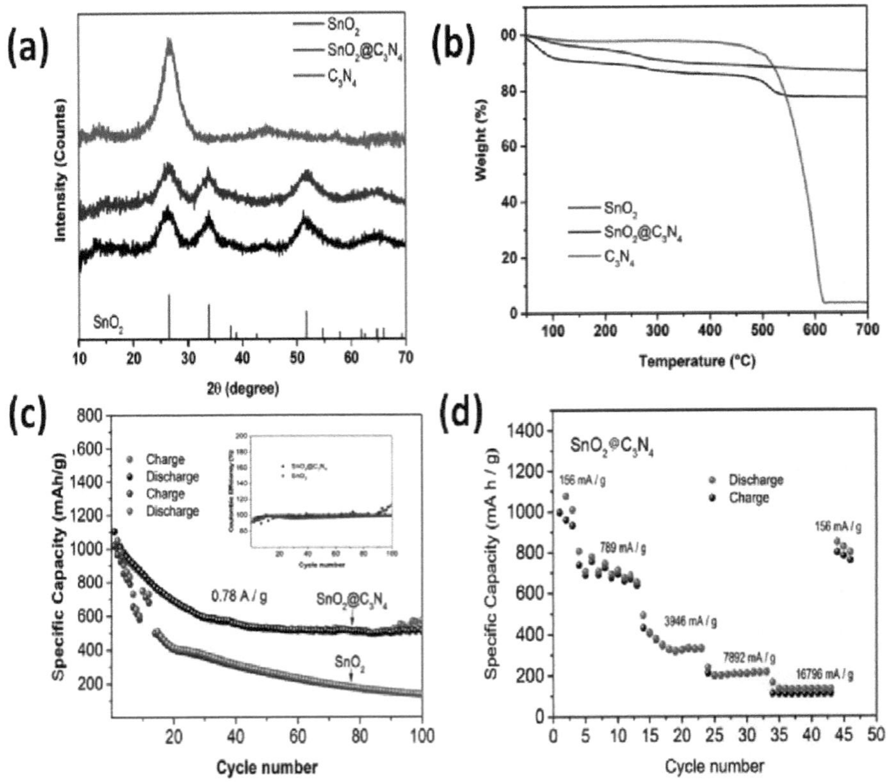

Figure 4.1 (a, b) X-ray diffraction patterns and TGA curves of g-C_3N_4, SnO_2, and SnO_2@g-C_3N_4. (c) Cyclic life performance of SnO_2 and SnO_2@g-C_3N_4 at 0.78 A/g. (d) Rate capability of SnO_2@g-C_3N_4 at different current densities. (Adapted from Ref. [34] Copyright 2020, *Solid State Ion.*, Elsevier).

The same strategy can be followed for different nanostructured metal chalcogenides to boost the electrochemical properties of LIBs. For example, Senthil and co-workers reported an increased reversible capacity of 857 mAh g^{-1} at a 0.1 C rate after 50 cycles for g-C_3N_4@MoS_2 scaffolds synthesized via the hydrothermal and sonication methods [36]. Current research has also been focused on the development of ternary g-C_3N_4-based composites, which can accommodate lattice strain and possess increased surface area for Li-ion intercalation and de-intercalation processes.

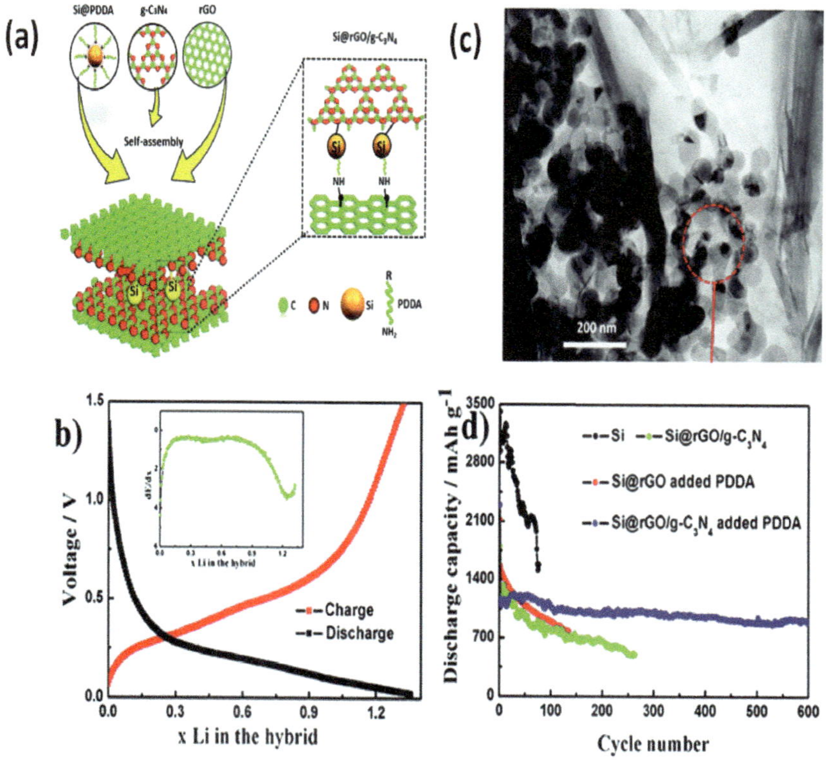

Figure 4.2 (a) Si@rGO/g-C_3N_4 schematic reaction scheme. (b) Si@rGO/g-C_3N_4 galvanostatic CD curve atsssw 0.5 C. (c) Si@rGO/g-C_3N_4 HR-TEM image. (d) Si, Si@rGO/g-C_3N_4 anode lifecycle performance at 0.2 C. (Adapted from Ref. [35] Copyright 2018, *J. Mater. Chem. A*, The Royal Society of Chemistry).

Hou et al. reported a 3D N-doped graphene (NG)/g-C_3N_4 nanosheet with layered MoS_2 [37]. They observed a high reversible capacity of 938 mAh g^{-1} with 91% charge efficiency after 100 cycles. The excellent performance was attributed to the N-rGO and porous g-C_3N_4 nanosheets, which played a role in buffering the volume fluctuations, facilitated electron and Li-ion transport, and prevented aggregation of MoS_2 nanosheets. Similarly, Yin et al. reported on SnS_2 coupled with g-C_3N_4 with 444 mAh g^{-1} at 100 mA g^{-1} after the 100th cycle [38]. Further, a few ternary hybrid nanocomposites such as rGO@SnS_2 with g-C_3N_4 nanosheets prepared by the hydrothermal method [39] or SnO_2-TiO_2 coated on g-C_3N_4 by the self-assembly deposition process as an anode material [40]. Both hybrid nanocomposites exhibit remarkable reversible capacity for Li-ion storage with a longer cycle life and excellent rate capability. Thus, various g-C_3N_4 nanocomposites were found to be performing electrode materials for LIBs.

4.3 g-C_3N_4 Electrodes for NIBs

Of late, the diminishing Li reserves and the increasing market requirement for high-energy batteries for electric vehicles have fueled research and development of NIBs. Being highly abundant on earth, NIBs can mitigate overdependence on Li, while also being a more affordable option with higher recyclability/recoverability. Although, the equivalent weight of Na is greater than Li and has direct implications on the derivable specific capacities (1.16 Ah g^{-1}), NIBs are more economically favorable than LIBs [41]. From a thermodynamic point of view, a comparable standard electrode potential for Na (–2.71 vs SHE), with lower desolvation of Na ions in organic solvents and lower charge transfer resistance, makes it a suitable replacement for Li. Similar to LIBs, during discharge/charge, Na ions diffuse from the anode to the cathode or vice-versa. However, the degree of intercalation for Na^+ varies, due to the larger ion size (1.16 Å) and leads to drawbacks like capacity fading and low cyclability, hindering its practical usage in batteries. The rapid interaction/de-intercalation of Na^+ ions during charge–discharge of NIBs leads to structural degradation of the anode. Thus, there is a need to develop suitable anode materials with a greater number of electrochemical active sites to accommodate Na ions and improve the specific capacity to achieve excellent battery performance.

Graphite anode used in LIBs can easily intercalate Li ions but the same cannot be expected for Na ions, due to its less covalent character and the electrochemically reversible nature of the Na^+. The insertion into graphite is hampered, owing to insufficient interlayer size (0.34 nm) to host larger Na^+ ions. Also, graphite electrode materials are thermodynamically unstable with high Na content and therefore suffer from very low Na storage (<35 mAh g^{-1}). According to theoretical and experimental studies, a slight expansion of the layer size to 0.37 nm (from

0.34 nm) can marginally reduce the energy barrier for Na^+ insertion into graphite [42]. Doping with heteroatoms can be a suitable approach to expand the layer size and effectively assist in altering the energy barrier for ion insertion. Analogous to graphite, g-C_3N_4 with nitrogen moieties can promote Na^+ insertion and its adsorption on N-moderated active sites, thus improving the electrochemical performance [43]. As such, g-C_3N_4 can serve as a promising anode candidate for NIBs, owing to easy scalability, low cost, chemical stability, and potentially high-rate capability. However, intrinsic g-C_3N_4 exhibits poor electronic conductivity, low reversible Na storage capacity, and insufficient cyclability.

In an attempt to lower the Na^+ ion diffusion barrier in g-C_3N_4 and boost electronic conductivity, Taylor et al. hypothesized and synthesized a C/g-C_3N_4 composite [44]. The composite could achieve two times higher Na^+ storage in NIB when compared to pristine g-C_3N_4 anode, owing to a multi-layered structure. In a half-cell configuration of the composite vs. Na^+/Na, the composite exhibits a specific capacity of 151 mAh g^{-1} at a discharge rate of 1 A g^{-1}. Moreover, an average specific capacity of 160 mAh g^{-1} was observed for over 400 cycles at 0.4 A g^{-1} for the half-cells. NIB fabricated using the composite as the anode and sodium rhodizonate dibasic ($Na_2C_2O_6$) as the positive electrode, exhibited discharge capacity as high as 120 mAh g^{-1} at 1 A g^{-1} for almost 14000 cycles. It has been found that layered materials contribute to fast Na-ion cycling, due to enhanced interlayer spacing. It is also important to note that further doping with sulfur or phosphorous can effectively boost the electrochemical activity. A DFT study on the electronic structure of P-doped g-C_3N_4 nanosheets for sodium storage was assessed, which depicted that the presence of the P atom causes the band gap of g-C_3N_4 to narrow, thereby improving conductivity [45].

Nevertheless, much research on improving the performance of g-C_3N_4-based electrodes for applications in full-fledged NIBs is required to be undertaken. The focus needs to be on tuning the interlayer space of such nanocomposites and asserting electronic conductivity along with structural stability. The use of transition metal oxides/sulfides or chalcogenides can be of significance in this perspective. Wang et al. reported facile synthesis of $Na_2Ti_3O_7$ nanotube/g-C_3N_4/graphene network and investigated electrochemical performance for NIBs. $Na_2Ti_3O_7$NT/g-C_3N_4/RGO composite exhibited a capacity of 210.8 mAh g^{-1} at a current density of 0.1 A g^{-1} even after 300 cycles with Coulombic efficiency more than 98% [46]. Similarly, Zhou et al. also successfully synthesized N-deficient g-C_3N_4 (ND-gC_3N_4) by magnesiothermic denitriding route and developed nanocomposites with MoS_2 as electrodes for NIBs [47]. The MoS_2-ND-g-C_3N_4 nanocomposite electrode delivered a high reversible specific capacity of 610.66 mAh g^{-1} at 0.1 A g^{-1}, superior rate performance, and remarkable long-cycle stability with 365.8 mAh g^{-1} at 1 A g^{-1} after 500 cycles for NIBs. Such superior performance has been due to the synergistic effect of ND-g-C_3N_4 and MoS_2 nanosheets. This design not only enhances the electrochemical kinetics of Na^+ diffusion mobility of MoS_2 but also leads to conductive pathways ushering in fast electron transfer. Other researchers have investigated various g-C_3N_4 nanocomposites, which depict future usage as anodes for NIB applications [48, 49].

4.4 g-C_3N_4-Based Electrodes for MABs

Metal–air batteries (MABs) have become sought-after energy storage devices for their high energy density (theoretically speaking) and high power-to-weight ratio, equally suitable for consumer electronics and automotive applications. The availability of ample resources, at the disposal, reusability, and reduced toxicity of the materials used in MABs, is a prime reason for encouraging rapid development and commercialization. One of the main attractions is convenient architecture, involving three main aspects: the oxidation of an active metal anode, reduction of air/ O_2 on a suitable cathode catalyst, and essential migration of ions mediated by a suitable electrolyte. However, large-scale distribution has been hindered due to the inadequate rate capabilities of the cathode catalyst for oxygen electrode reactions involving multiple proton-electron transfers at a triple-phase boundary (solid catalyst, liquid electrolyte, and gaseous oxygen). Complete ORR/OER on an ideal catalyst, in a suitable electrolyte, follows a $4e^-$ transfer process. On most catalysts, reactions take a setback due to a large energy barrier for O_2 dissociation/

formation, which invariably leads to alternate pathways and generates stable intermediates that block active sites. Tackling oxygen electrocatalysis effectively involves the use of catalysts, which possess good electronic conductivity for efficient electron transfer, low O_2 dissociation/formation energy barrier, and durability. The above attributes are usually associated with catalysts based on rare-earth metals such as Pt, Ir, and Ru. The adoption of these catalysts dramatically increases the costs of production, due to their scarcity and high cost. This prompts research for stable electrocatalysts that can be generated from earth-abundant elements and can be equally competent to precious catalysts for OER and ORR.

Among non-precious electrocatalysts that can be synthesized by cost-effective and eco-friendly methods, with significant $4e^-$ oxygen activity, electrical conductivity, and chemical resistance, g-C_3N_4 has gathered much attention. What sets it apart from conventional carbo-nitro compounds is their polymeric nature, thermo-chemical stability, high surface area, and most of all, electron-rich properties, owing to the presence of N lone-pair electrons, rendering them, as sites for metal inclusion and electronic modification of molecular structures [50]. Although it has a comparable onset potential to that of a Pt/C catalyst and is capable of selectively following a $4e^-$ ORR pathway [51], g-C_3N_4 falls short of the requisite catalytic activity desirable for ORR applications, owing to its low charge transferability. The ORR sites on pure g-C_3N_4 have limited coverage at the electrode-electrolyte-gas triple-phase boundaries (TPB) because of poor electrical conductivity (10^{-9} S m^{-1}) [52]. Theoretical considerations on the mechanistic behavior of the ORR on g-C_3N_4 by Zheng et al. reveal that the low charge transfer ability of the carbon nitride leads to the accumulation of the OOH^- intermediate via an inefficient and incomplete $2e^-$ pathway [53]. To increase the charge and transfer efficiency, the authors incorporated g-C_3N_4 into the mesoporous CMK-3 carbon framework, achieving a 100% $4e^-$ pathway for ORR. The ORR activity was enhanced due to the CMK-3 carbon framework, which extends conductive support for homogenously distributing the g-C_3N_4 and ensures an effective charge transfer.

The role of a conductive component to increase the charge transfer efficiency and ORR activity can certainly be fulfilled by transition metals such as Ni, Cu, Co, Fe, and Mo, or a combination of them, due to their inherent electrical conductivity and affinity for adsorption of O species. In addition to that, like g-C_3N_4, the transition metals also possess electrochemical stability and durability in varying ranges depending on their composition. Density functional theory (DFT) investigations by Chen and Hu have identified M-OH-C_3N_4 metal complexes (M = Ni or Cu) as stable ORR electrocatalysts with activities outperforming even that of Pt(111) [54]. Experimentally, g-C_3N_4-based catalysts integrated with transition metals such as Co [55], Cu [56], Ag [57], and Ce [58] have been reported to be active for ORR. The integration of Co with the conductive graphene and g-C_3N_4 was sought by Liu and Zhang by developing a nitrogen-metal macrocyclic ORR catalyst [55]. The Co/g-C_3N_4@G composite catalyst was capable of onset potential and current density comparable to that of Pt/C, with active sites favoring 4e^- oxygen reduction. Besides chronoamperometric discharge for 10 h, at −0.45 V (vs. Ag/AgCl), the catalyst also exhibited superior stability than Pt/C. The excellent ORR catalytic activity has been assigned to the formation of active Co-N_x moieties, which serve as active sites and enhance charge transfer efficiency in the composite catalyst. Macrocyclic Cu-N_x moieties with pyridinic N in Cu@g-C_3N_4 catalyst also exhibit low peroxide formation and low onset potential with Tafel slope, much lower than that of even Pt/C, due to the weakening of the O–O bond adsorbed at the CuN_x sites [56]. The introduction of the highly conductive Ag into g-C_3N_4, by Bhandaray et al. also displayed a drastic improvement in the exchange current densities for ORR by three orders of magnitude [57].

Oxygen evolution reaction also forms an important part of oxygen electrocatalysis and it is desirable that catalysts exhibit a bifunctional activity for OER and ORR. Fortunately, owing to the presence of pyridinic-N sites, several g-C_3N_4-based catalysts also have the potential to be efficient catalysts for OER, as well. For instance, g-C_3N_4 nanosheet composites by Tian et al. showcase significant activity and stability toward OER, comparable to that of RuO_2, with a low Tafel slope of 68.5 mV dec^{-1} and low overpotential (η_{10} = 539 mV) [59]. As in the case of ORR, OER on g-C_3N_4 is also limited by charge transfer efficiency and lack of active sites. To increase the OER activity and stability of g-C_3N_4 electrodes, integrating with transition metals, oxides, hydroxides, heteroatom-doping, chalcogenides, etc. has been adopted.

Following the advancements in the O_2 electrocatalytic behavior of g-C_3N_4-based catalysts, the application of the same is lucrative in MABs. However, owing to different battery chemistries in aprotic and parotic solvents, the activity of a particular catalyst may vary. For instance, the reactive nature of Li or Na metal restricts the use of aqueous solvents and Li–O_2/Na–O_2 batteries typically involve an organic electrolyte. Contrary to this, MABs that employ metals non-reactive to water such as Zn, Mg, and Fe are free to adopt aqueous electrolytes. In all cases, the ORR/OER activity of the catalyst is influenced by the chemical/electrochemical interaction of the electrolyte molecules or generated intermediates, reflecting the wide available choice for organic solvents, as evident from the review by Balaish et al. on Li-electrolytes [60]. And the complexities of the ORR/OER mechanisms for various electrolytes, we have restricted discussion only to the electrochemical performance of Li–O_2 batteries, irrespective of the electrolyte used [61].

4.4.1 g-C_3N_4-Based Electrodes in Li–O_2 Batteries

High theoretical specific energy of 5200 Wh kg^{-1} (including O_2) renders Li–O_2 batteries as promising energy storage devices, capable of ushering in significant advances in the field like the LIBs. Li–O_2 cell is composed of a Li-metal anode, Li^+ conducting electrolyte, and porous air electrode as the cathode. Due to the reactive nature of Li metal to moisture, a non-aqueous electrolyte is employed in Li–O_2 batteries. According to the overall equation:

$$2\,Li^+ + O_2 + 2\,e^- \leftrightarrow Li_2O_2. \tag{4.1}$$

During discharge, ORR at the cathode produces Li_2O_2, while charging, it leads to the OER, decomposing Li_2O_2 to regenerate O_2 and Li^+. The insolubility and poor conductivity of Li_2O_2 result in energy losses, arising from increasing charging voltages required for Li_2O_2 decomposition in practical Li–O_2 cells, compared to the standard thermodynamic potential for solid Li_2O_2 formation and decomposition (2.96 V vs Li^+/Li). Additionally, the formation of Li–O_2 from superoxide species (O_2^-) generated during ORR, which is irreversible with respect to Li^+/Li, also poses challenges for the effective rechargeability of the Li–O_2 cells. The Li–O_2 battery chemistry differs from the rest of the MABs, due to the different O_2 electrocatalytic mechanisms in non-aqueous electrolyte systems.

Owing to the O_2 reversible reaction capabilities of g-C_3N_4 and stability in organic solvents, several g-C_3N_4-based catalysts have been adopted in Li–O_2 batteries. The introduction of graphene as a conductive support for g-C_3N_4 (G@CN) by Lou et al. resulted in a free-standing, binder-free, and metal-free electrocatalyst with excellent rechargeability and round-trip efficiency for OER and ORR in LABs [61]. The Li–O_2 cell with the composite demonstrated stability for more than 100 cycles with discharge potentials higher than 2.7 V and charging potentials lower than 4.4 V (Fig. 4.3b). The presence of a large number of nucleation sites and the superior electrical conductivity of the composite are responsible for the enhanced cycling stability of the G@CN-based Li–O_2 cells. The composite catalyst also demonstrated enhanced OER and ORR activities at small overpotentials and $4e^-$ pathway, indicating the suitability for aqueous electrolytes, as well. Following inspiration from the above work, Yi et al. achieved g-C_3N_4 integrated on carbon paper (CP) by in situ deposition and fabricated a Li–O_2 cell from the g-C_3N_4@CP as the ORR cathode [62]. The battery attained a high discharge capacity of 1000 mAh g^{-1} when discharged at different current densities. Capacity-controlled discharge at 300 mAh g^{-1} for the Li–O_2 cells showcased negligible decay for more than 100 cycles at a current density of 500 mA g^{-1}, indicating excellent cycling stability (Fig. 4.3d). Definitely, the integration of g-C_3N_4 with carbon seems to be a low-cost and eco-friendly approach for efficient cathodes for Li–O_2 cells. However, Li–O_2 cells based on such catalysts, may not adequately meet the energy demand, unless O_2 electrocatalysis is improved.

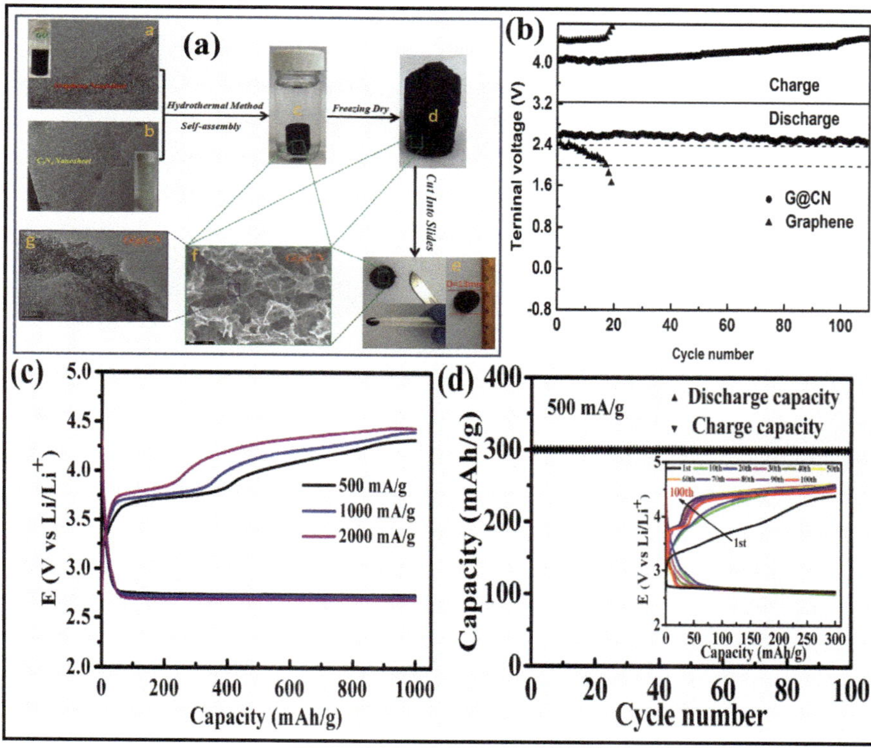

Figure 4.3 (a) Schematic flow-process diagram, HRTEM images of the G@CN composite. (b) Comparison of the cycling performances of graphene and G@CN free-standing macroporous electrode. (Reproduced with permission from Ref. [61]. Copyright 2015, *Small*, WILEY-VCH.). (c) Rate capability of g-C_3N_4@CP at different current densities when discharged to 1000 mAh/g. (d) Cycling performance and the corresponding discharge/charge profiles (inset) of g-C_3N_4@CP at the current density of 500 mA/g with a capacity limit of 300 mAh/g. (Reproduced with permission from Ref. [62]. Copyright 2015, *Appl. Mater. Inter.* American Chemical Society).

The decomposition of the Li_2O_2 is a major factor limiting the rechargeability and energy efficiency of the Li–O_2 batteries and different strategies have been adopted to tackle this. One of the approaches involves the use of an active electrocatalyst to lower the overpotential for Li_2O_2 decomposition. The combination of semi-conducting $CoSe_2$ and g-C_3N_4 by Kumar et al. resulted in decreasing the charge voltage by 280 mV, while having no appreciable voltage change on the discharge plateau [63]. Even though the cell retained 100% CE, the discharge capacity decreased after seven cycles, either due to the degradation of the DMSO electrolyte or the destabilization of the ORR catalyst. The cells exhibited better cyclic stability in ether-based (TEGDME) solvent, the exact reason for which is not discerned. Integration of α-MnO_2 nanorods on g-C_3N_4 was carried out hydrothermally by Hang et al. and was found to exhibit excellent OER and ORR activities [64]. Li–O_2 cells with α-MnO_2/g-C_3N_4 demonstrated enhanced performance compared to that of α-MnO_2 when evaluated over 40 cycles at a current density of 500 mA g^{-1} and a controlled capacity of 1000 mAh g^{-1}. The authors attribute the improved performance to the porous g-C_3N_4 structure and the surface mobility of peroxide species on α-MnO_2, leading to the inhibition of Li–O_2 formation while charging and the decomposition of Li_2O_2 species, during discharge. The potential enhancement of O_2 electrocatalysis following the incorporation of perovskite materials into g-C_3N_4 composites was demonstrated by Wu et al. [65]. $LaNiO_3$-g-C_3N_4 composite was used as an ORR cathode in Li–O_2 cells, resulting in a low charge voltage gap of 1.2 V and a discharge plateau of 2.8 V. The cyclic stability of the composite extended beyond the normally reported values, proceeding to 65 cycles for a discharge capacity of 250 mAh g^{-1}. In a significant achievement, Li–O_2 cells using boron-doped C_3N_4 with RuO_2 composite cathode were found to exhibit excellent rechargeability and a very low voltage gap of ~0.8 V [66]. The enhanced performance of the composite has been attributed to the structural stability offered by the B-doped g-C_3N_4 and the improved ORR/OER activity, due to the incorporation of RuO_2 nanoparticles. The authors also note the inhibition of CO_3^{2-} formation in the electrolytes, due to the absence of the 'labile' carbon in the composite.

Another approach to decrease the voltage gap, which gathered recent attention is the use of a redox-mediator (RM) in the liquid phase, along with the air electrode. To some extent, the introduction of RMs as dissolved charge transfer entities has been successful in reducing the voltage gap, by catalyzing Li_2O_2 decomposition.

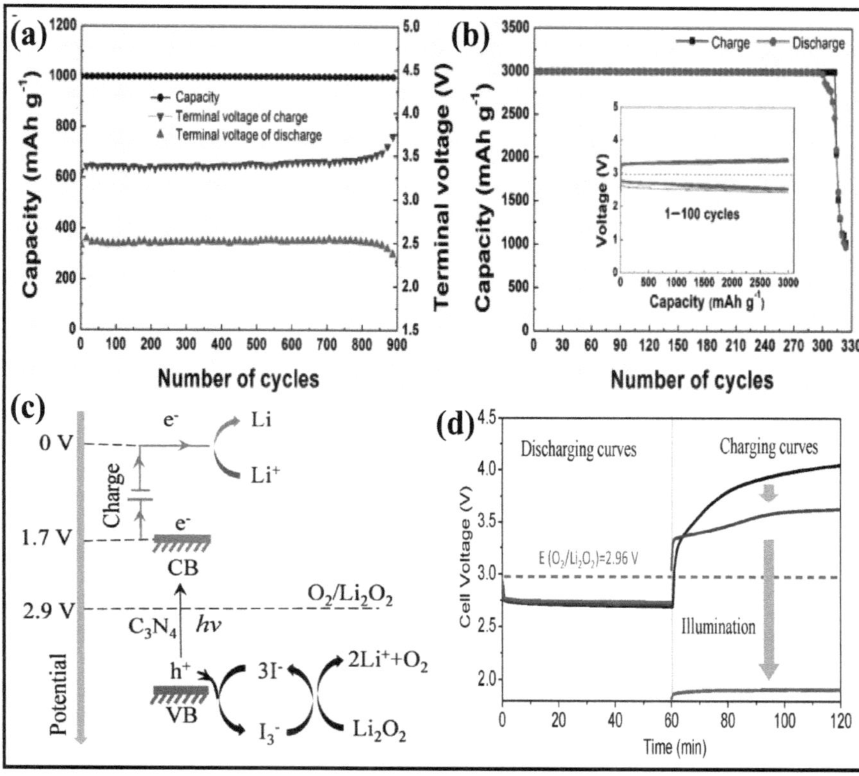

Figure 4.4 (a) Electrochemical profiles (b) cyclability and terminal voltages of the CNT fibril electrodes with the LiI RM; cyclability of the CNT fibril electrodes in the presence of a LiI RM at a discharge depth of 3000 mAh g^{-1}. (Reproduced with permission from Ref. [68]. Copyright 2014, *Angew. Chem.*, Wiley-VCH). (c) In a photo-assisted charging process, the photoexcited electrons of g-C_3N_4 transfer to the anode to reduce Li^+ to Li metal, and the generated photovoltage is utilized to compensate for the required charging voltage. The photo-assisted charging voltage equals the energy difference between the redox potential of the Li^+/Li couple and the CB of a photocatalyst (1.7 V). Meanwhile, the redox mediator I^- ions are oxidized to I_3^- by the photoexcited holes and then, in turn, oxidizes the discharge product Li_2O_2; in the process, I_3^- is reduced back to I^-. (d) The charging/discharging curves of a Li–O_2 battery without the I^- ion redox mediator (black line), a Li–O_2 battery with the I-ion redox mediator (blue line) and the photo-assisted rechargeable Li–O_2 battery (red line). (Reproduced from Ref. [69] with permission from The Royal Society of Chemistry. Copyright 2015, *Energ. Environ. Sci.*, The Royal Society of Chemistry).

The RM should have an oxidation potential higher than that of the Li_2O_2 equilibrium potential (2.96 V vs Li^+/Li), it is active only while charging, and parasitic reactions can be avoided during the discharge process as given in Fig. 4.4. Apart from being able to efficiently catalyze Li_2O_2 decomposition, RMs should be unreactive with the electrolyte or Li anode. An exhaustive review of the RMs adopted in Li–O_2 batteries has been done by Park et al. [67]. Kang and co-workers were able to decrease the charging voltage to 3.5 V, for Li–O_2 cells with LiI RM and aligned fibrous-CNT catalysts, with rechargeability up to 900 cycles at a discharge capacity of 1000 mAh g^{-1} and a current rate of 2000 mA g^{-1} [68]. The Li–O_2 cells were capable of performing for 300 charge/discharge cycles even on increasing the discharge capacity, by three-fold. In conjunction with the redox-mediated process, a photo-assisted charging approach involving a g-C_3N_4 photocatalyst was adopted by Liu et al. [69]. Quite surprisingly, the new configuration was able to achieve an energy efficiency of 142%, with a photo-assisted charging voltage of 1.9 V, which is lower than even the discharge voltage. The photovoltage generated by the g-C_3N_4 compensates for the charging overvoltage and accelerates Li_2O_2 decomposition by enhancing charge transfer, contributing to Li^+ reduction and oxidation of I^-.

Different strategies and approaches, focusing on key attributes such as porosity, charge transfer efficiency, ORR/OER capabilities, and stability against electrolyte solvents for efficient O_2 electrocatalysis have to be borne in mind for efficient Li–O_2 cells. From an air-electrode point of view, the integration of g-C_3N_4 with active ORR/OER electrocatalyst components will benefit from reducing the active catalyst mass while also ensuring a porous distribution platform for active sites. Overall, the shortcomings of high overpotential losses, kinetic limitations, inefficient rechargeability, and cycling stability have been attempted to solve, by present-day g-C_3N_4 catalysts. However, a significant amount of improvement in catalyst development is required to realize commercial rechargeable Li–O_2 cells. In this context, the existing literature and developmental research on g-C_3N_4 definitely pave the base for building up efficient and effective catalysts.

4.4.2 g-C_3N_4-Based Electrodes in ZABs

Zinc–air batteries (ZABs) have garnered attention, owing to the abundance of Zn, durability of Zn in aqueous solutions, low environmental concerns, reduced safety hazards, and scope for cost-effective and scalable manufacture. Analogous to the architecture of the LABs, ZABs are composed of a Zn anode, an aqueous electrolyte, and an air cathode. During the discharge process, the oxidation of the Zn anode and the reduction of O_2 occur at the cathode catalyst. Though having a theoretical voltage of 1.667 V and a specific energy density (theoretical) of 1.08 kWh kg^{-1}, due to sluggish reaction kinetics, ZABs exhibit a low practical open-circuit voltage (OCV) of 1.35 V.

$$\text{Anode: Zn} + 4\ \text{OH}^- \rightarrow \text{Zn(OH)}_4^{2-} + 2\ \text{e}^-\ (E_0 = -1.25\ \text{V}) \qquad (4.2)$$

$$\text{Cathode: } 1/2\ \text{O}_2 + \text{H}_2\text{O} + 2\ \text{e}^- \rightarrow 2\ \text{OH}^-\ (E_0 = 0.34\ \text{V, pH 11}) \qquad (4.3)$$

$$\text{Overall: } 2\ \text{Zn} + \text{O}_2 \rightarrow 2\ \text{ZnO}\ (E_0 = 1.59\ \text{V}) \qquad (4.4)$$

Unlike Li–O_2 batteries, ZABs do not pose fire hazards and do not require special equipment such as glove boxes for fabrication, which is exactly why they are referred to as 'zinc–air' batteries and not as 'Zn–O_2'. Apart from that, the advent of ZABs encourages the recyclability and reusability of discarded anodes from primary batteries after mild chemical treatment. Despite these benefits, just as in the case of Li–O_2 cells, ZABs also face overpotential and rechargeability issues, owing to the sluggish ORR/OER kinetics. Another concern for ZABs is the HER at the Zn anode leading to self-discharge and degradation issues. Although primary ZABs have been used for quite a long time, such as in hearing aids, seismic telemetry, and railroad signaling, only in recent years, electrochemically rechargeable ZABs have been realized [70].

Graphitic carbon nitride–carbon nanofiber composites were effectively employed by Sung et al. as air cathodes in primary and rechargeable ZABs and were found to exceed the performance of ZABs with Pt/C cathode, at the same loading [71]. The improved electrical conductivity, corrosion resistance, and sufficient active sites of the composite for reversible O_2 catalysis significantly contribute to the improved performance. Heteroatom-doping in g-C_3N_4 is an important strategy for the development of efficient electrocatalysts while also ensuring ease of synthesis and large-scale production, and can effectively increase the power-to-weight of ZABs. Doping of P and S heteroatoms in g-C_3N_4 sponges was adopted by Shinde et al. to enhance the electrical conductivity of the g-C_3N_4 catalyst for rechargeable ZABs, resulting in outstanding bifunctional O_2 electrocatalysis [72]. In terms of ORR/OER activities in alkaline aqueous solutions, the P- and S-doped g-C_3N_4 hybrid catalysts showed better electrocatalytic activity than Pt/C (for ORR) or RuO_2 (for OER). Consequently, the application of the catalyst in primary ZABs resulted in a high PD of ~198 mW cm^{-2}, an OCV of 1.51 V, and a high specific capacity of 779 mAh g^{-1} at a discharge CD of 20 mA cm^{-2} for more than 4 h. Rechargeable ZABs using the hybrid catalyst demonstrated 200 cycles over 40 h at a CD of 2 mA cm^{-2}. Interestingly, a tri-electrode configuration of the bifunctional catalyst (separate electrodes for the OER and ORR) resulted in outstanding stability for 500 discharge/charge cycles over 100 h at CDs of 25 mA cm^{-2}, better than even the Pt/RuO_2 couple for ORR/OER, respectively. Phosphorous- and sulfur-doped carbon nitride fibers (P- and S-CNF) on carbon cloth, developed by the same group using the polymerization synthesis, also exhibited outstanding activity in rechargeable ZABs [73]. The ZABs fabricated with P- and S-CNF catalysts demonstrated durability and stability over 600 discharge/charge cycles over 120 h at a CD of 2 mA cm^{-2}. Flexible rechargeable solid-state ZABs were also reported using the hybrid P- and S- CNF catalysts, showcasing a discharge voltage of 1.1 V for 9 h at a current density of 2 mA cm^{-2}. High specific surface area, mesoporous structure, and electronic modifications introduced in the carbon nitrides, due to the electronegative heteroatoms, have been attributed to the excellent performance of the hybrid catalysts. Rechargeable ZAB using CoS_x@g-C_3N_4/rGO catalyst was also reported by Yang et al. to exhibit cyclic stability for 394 cycles over ~43 h, better than even the 171 cycles over 9 h demonstrated by commercial Pt/C as shown in Fig. 4.5 [74]. The porous structure, improved electrical conductivity, and accessible nitrogen sites for ORR/OER are responsible for the rechargeability and durable electrocatalysis. There is still scope for improvement in the field of ZABs and g-C_3N_4 can be a reliant platform for the synthesis of high-performance bifunctional catalysts.

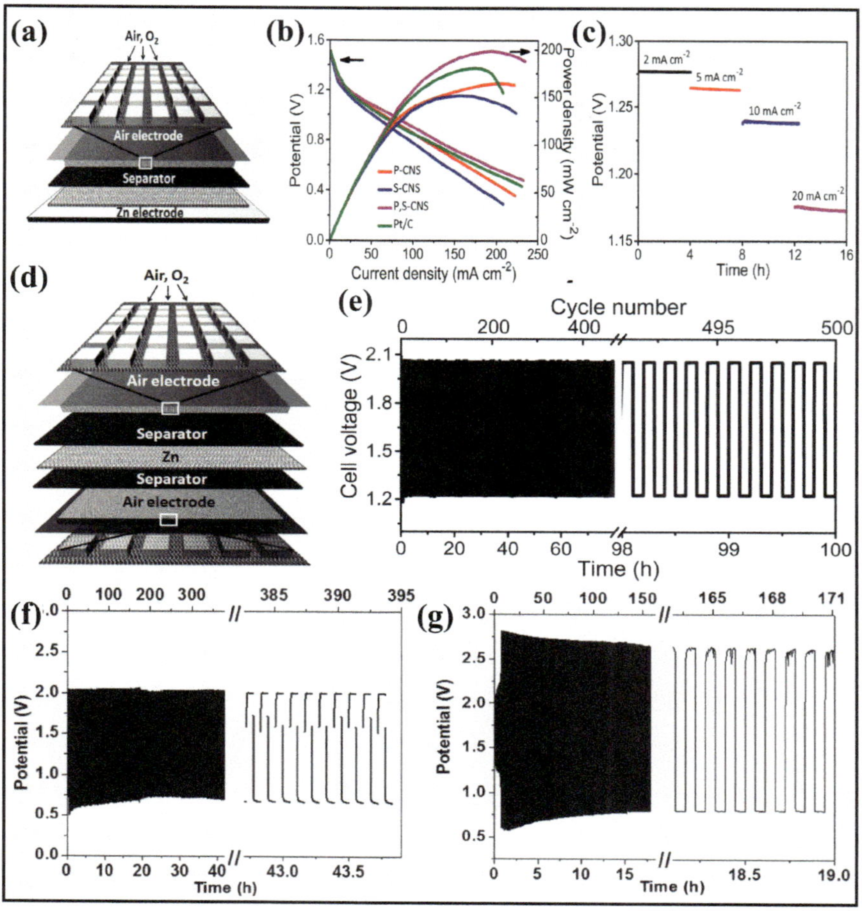

Figure 4.5 (a) Schematic representation for the primary ZABs. (b) The primary ZABs power density and polarization plots with various catalysts. (c) Galvanostatic discharge plots with P- and S-CNS air-cathode for the primary ZABs at various current densities (regularized with the vicinity for the air-electrode). (d) Schematic illustrations of tri-electrode ZABs. (e) Three-electrode rechargeable ZABs galvanostatic charge-discharge cycles with P- and S-CNS as the bifunctional catalyst. (Reproduced with permission from Ref. [72]. Copyright 2017, *Nano*, American Chemical Society). Galvanostatic discharge/charge cycling curves of (f) CoS_x@PCN/rGO and (g) Pt/C+RuO_2 in rechargeable ZAB at a current of 50 mA. (Reproduced with permission from Ref. [74]. Copyright 2017, *Adv. Energy Mater.*, Wiley-VCH).

4.4.3 g-C_3N_4 Electrodes in Other MABs

Apart from the above, other prominent candidates in the realm of MABs that have already gathered significant attention would be aluminum–air batteries, magnesium–air batteries, and iron–air batteries. Such MABs have promising theoretical EDs, wide availability, and low cost of the anode materials. Foraging from whatever little has been published in literature until now, there is a definite improvement in performance and scope for these MABs based on g-C_3N_4 catalysts. For instance, rechargeable aluminum–air batteries fabricated using Co-g-C_3N_4 catalyst for over 22 cycles at 10 mA cm^{-2}, faring better than the four cycles achieved by commercial Pt/C + RuO_2/C hybrid catalysts, with a 200 mV lower voltage gap [75]. Thus, scientific advancements in the field of O_2 electrocatalysis, heralded by g-C_3N_4-based catalysts, can prove significant for the commercialization of MABs.

4.5 g-C_3N_4-Based Li–S Batteries

Lithium–sulfur (Li–S) batteries have garnered attention in recent years, as low-cost energy storage devices. Structurally similar to the Li–O_2 batteries, Li–S batteries involve the use of a Li-metal anode, an elemental-sulfur containing cathode, and a compatible electrolyte, resulting in a theoretical capacity of 1672 mAh g^{-1} and an energy density of 2567 Wh kg^{-1}. At the anode, the Li metal is oxidized to Li^+, following the reaction:

$$16\ Li \leftrightarrow 16\ Li^+ + 16\ e^- \qquad (4.5)$$

and reduction of sulfur to Li_2S occurs at the cathode, according to the reaction:

$$S_8 + 16\ Li^+ + 16\ e^- \leftrightarrow 8\ Li_2S\ (E^o \approx 2.15\ V\ vs\ Li/Li^+). \qquad (4.6)$$

Despite the high theoretical energy densities and capacities, the practically achievable energy density of Li–S batteries is limited to the low conductivity of the cathode, insulating properties of the Li_2S, low loading of S and high dissolution of Li polysulfide

(LiPS) intermediates (shuttle effect), resulting in an irreversible loss of sulfur. Decay in electrochemical performance, owing to the shuttling effect is a great challenge that should be overcome to tap the full potential of Li–S batteries.

The cathode optimization is an importantly adopted approach to improve Li–S battery performances, focusing on a large electrochemical surface area to increase S loading. Following this, some of the early sulfur cathodes involved the use of high surface area carbon-based materials such as mesoporous carbon [75–78], hollow carbon nanospheres [79–81], and graphene [82, 83]. Though, the cathodes provide ample surface area for the adsorption of LiPS intermediates, and thereby prevent the 'shuttle effect', the rather weak physical interaction between the non-polar carbon and polar LiPS is not effective in suppressing the slow dissolution of the intermediates. So, the focus shifted to employing low-weight polar cathodes, which can inhibit LiPS dissolution and avoid a decrease in the energy density of Li–S batteries. In tune with this, utilizing the intrinsic charge polarity of the g-C_3N_4 for binding LiPS intermediates was proposed by Pang and Nazar [84]. The polymeric nature of the g-C_3N_4 can cater to the volume expansion of the cathode by Li_2S upon discharge. Theoretical calculations by Pang and Nazar also revealed higher binding energy for Li_2S on g-C_3N_4 than N-doped carbon, due to a greater number of pyridine sites. In terms of performance, Li–S batteries fabricated using 75 wt.% S and g-C_3N_4 retained 600 mAh g^{-1} capacity after 500 cycles at a fading rate of 0.06% per cycle, when examined at a 0.2 C-rate. However, after a graphene additive was introduced in the slurry the performance of the Li–S battery increased, more likely due to the conductive pathway offered by graphene, which cannot be offered by g-C_3N_4. In an effort to combat the 'shuttle effect' and low conductivity, Gong et al. synthesized porous g-C_3N_4@rGO (PCN@rGO-s) as sulfur cathodes for Li–S batteries [85]. From an initial capacity of 986 mAh g^{-1} at a 0.5 C-rate, the capacity of the Li–S batteries (sulfur loading of 4 mg cm^{-2}) decayed at the rate of 0.048% per cycle, up to 680 mAh g^{-1} after 800 cycles, indicative of the cycling stability of the cathodes. The authors assert strong chemical interaction of the PCN@rGO-s cathode and LiPS intermediates, increased conductivity due to the 3D rGO network, and the availability of open channels for ion diffusion, are the reasons for the long cycle life and high energy density.

Lithium polysulfide trapping leads to a decrease in Li^+ diffusion in the bulk of the cathode and results in low electrochemical performance of the Li–S batteries, which sought the need for a dedicated LiPS trap. Interlayers between the cathode and separator adopted in Li–S batteries serve the function without compromising cycle life. These cater to minimize the shuttle effect by acting as traps to dissolve polysulfide species via physical or chemical interactions. Although the adoption of a carbon paper interlayer to separate the cathode and separator by Manthiram et al. [86] was one of the pioneering works using this approach, the advent of g-C_3N_4-based interlayers significantly improved Li–S battery performance and cycle life. DFT calculations by Zhang et al. point to stronger binding of Li_2S to g-C_3N_4 interlayer compared to C-coated cathodes [87]. In the same work, the superior electrochemical performance of the Li–S batteries with g-C_3N_4 interlayer and carbon black cathode was demonstrated to have a capacity of ~730 mAh g^{-1} after being discharged at 1 C for 500 cycles. Similarly, Dai et al. successfully improved the electrochemical performance of Li–S batteries using laminar structured g-C_3N_4 and graphene composite as an interlayer for S-filled Ketjen black cathode shown in Fig. 4.6 [88]. The Li–S battery fabricated was capable of delivering a capacity as high as 612.4 mAh g^{-1} after 1000 cycles at a C-rate of 1 C. The introduction of the g-C_3N_4/graphene interlayer suppressed the shuttle effect by chemically anchoring long-chain polysulfides and at the same time, allowed for efficient Li^+ diffusion in the cathode.

Though g-C_3N_4-based interlayers effectively suppress the shuttle effect, the reduction of LiS_x to elemental S is also important for long cycle life and capacity retention in Li–S batteries. In an attempt to achieve a combined strategy for S redox reaction, improve Li^+ diffusion, and reduce the shuttle effect, Dong et al. proposed the use of a sandwich structure of g-C_3N_4, BN, and graphene sheets [90]. The fabricated Li–S battery benefited from such a configuration, owing to the synergistic effect of the individual 2D materials. The g-C_3N_4 and BN sheets catalyzed redox reactions of trapped LiPS intermediates and the graphene sheets served to improve the electronic conductivity of the interlayer. Although, Li–S battery with a BN/graphene interlayer

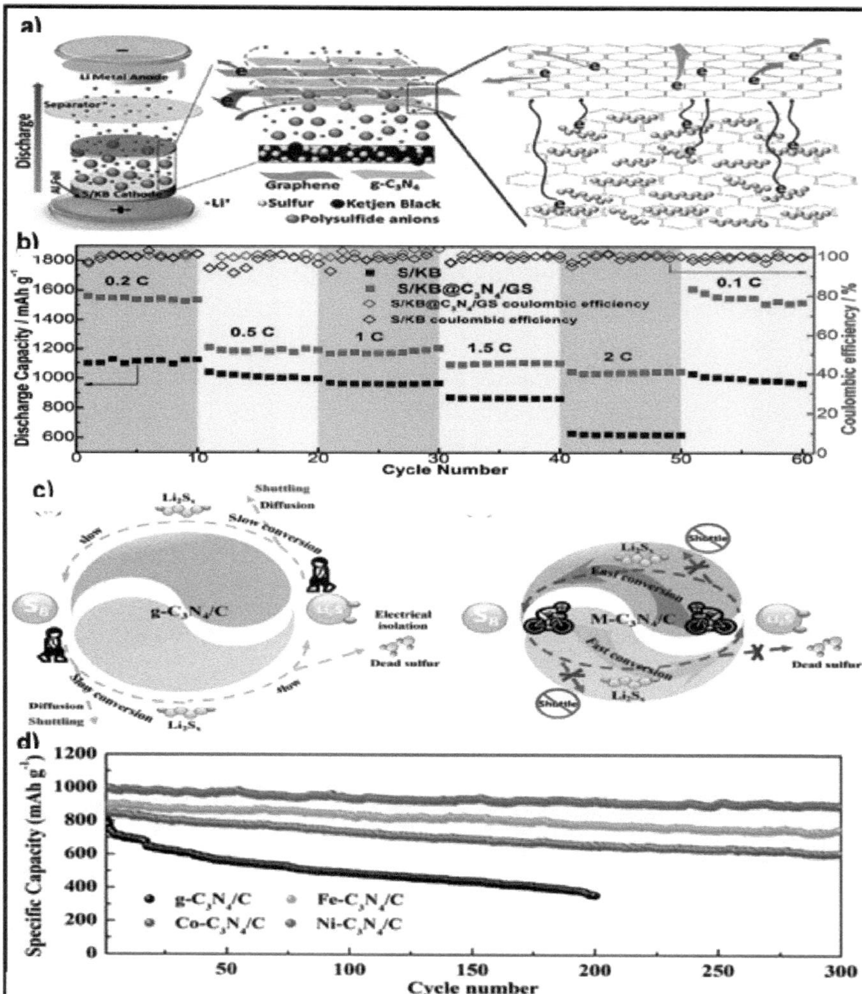

Figure 4.6 (a) Schematic of cell configuration with a laminated structure g-C_3N_4/GS cathode interlayer, and (b) Discharge capacities continuously cycled at various rates. (Adapted with permission from Ref. [88]). Copyright 2019, *ChemSusChem*, Wiley-VCH. (c) Schematic representation of the polysulfide conversion process on g-C_3N_4/C and M-C_3N_4/C, and (d) Charge-discharge curves of the Li–S batteries with the Ni-C_3N_4/C-modified separator at cycling performance at 0.5 A g^{-1}. (Adapted with permission from Ref. [89] Copyright 2019, *Chem. Eng. J.*, Elsevier).

and low sulfur loading (1.5 mg cm^{-2}) exhibited a high capacity of ~800 mAh g^{-1} for 300 cycles at a discharge rate of 1 C, higher S loading (6 mg cm^{-2}) resulted in decreased capacity. At high S loading, the large pore size (4 nm) of the BN-graphene layer does not effectively limit the shuttle effect.

However, the Li–S battery with a g-C_3N_4/graphene/BN sheets interlayer and S loading of 6 mg cm^{-2} demonstrated high cyclability (over 500 cycles) with a capacity retention of ~ 600 mAh g^{-1} at 1 C, with a decay rate of 0.01% per cycle. Transition metal centers coordinated with g-C_3N_4 structures have also been shown to be effective in inhibiting the shuttle effect and accelerating polysulfide conversion kinetics. Among the adopted Ni-, Fe-, and Co-based, Li–S battery fabricated using the Ni@g-C_3N_4 interlayer by Wang et al. exhibit capacity as high as 893 mAh g^{-1} at a current density of 0.5 A g^{-1} even after 300 cycles [89]. DFT calculations reveal that the dissociation energy of Li_2S on Ni-modified g-C_3N_4 is quite low (0.64 eV), offering catalytically active sites for Li_2S decomposition. Moreover, based on DFT studies, authors claim that the introduction of Ni into g-C_3N_4 extends a metallic character to it and thereby, effectively increases the electronic conductivity of g-C_3N_4.

4.6 g-C_3N_4-Based Supercapacitors

In the context that batteries are incapable of generating high power densities, it becomes imperative to work on the development of supercapacitors, which can complement the batteries in energy-intensive applications. Basically, a supercapacitor consists of two electrodes in an electrolyte, and electric charge is stored in the electrodes, which is then disbursed upon connecting to an external load. Based on the type of electrode materials used, the SCs are categorized into electric double-layer capacitors (EDLCs), pseudocapacitors, and battery-type supercapacitors. The and pseudocapacitors follow charge storage mechanisms at the active electrode surface by a non-Faradaic process, while the battery-type electrodes undergo reversible redox reactions, similar to a battery material [91]. EDLCs have high charge/

discharge efficiency, high power density, long cycling stability, and typically involve carbon-based electrodes [92]. But carbon-based materials are limited in specific capacitance, due to the physical surface charge accumulation that occurs at the electrode/electrolyte interface [93]. Carbon-based materials when doped with heteroatoms such as oxygen, sulfur, boron, nitrogen, and phosphorous have proven to be effective electrodes in EDLCs as heteroatoms enhance the electron mobility and capacity of charge storage [94]. Among these, nitrogen doping has been exploited as a means to enhance electrochemical active sites, electronic conductivity, and polarity. The amount of nitrogen content doped in the carbon matrix can be employed to tailor the wettability of the electrode material toward the electrolyte, thereby affecting diffusion and charge storage [95]. Out of two different configurations of nitrogen in nitrogen-doped carbon species viz. non-graphitic and graphitic, only non-graphitic nitrogen, i.e., pyridinic-N and pyrrolic-N offer capacitive performance. Recently, g-C_3N_4 has become a material of choice for EDLCs, due to its layered structure of carbon sheets with high nitrogen content [96]. Owing to the higher surface area electronic properties and a unique structure analogous to graphite, g-C_3N_4-based materials are exceedingly suitable for SC applications. The g-C_3N_4 exhibits improved electron donor properties, surface wettability, and chemical stability, which make it competently suitable for use in a wide range of pH. In addition to that, being synthesized from earth-abundant components renders it a safe and environment-friendly electrode material [49]. The g-C_3N_4 acts as a matrix that can provide mechanical and chemical stability and can effectively incorporate different materials to form nanocomposites. In this section, we will look into the supercapacitor applications of g-C_3N_4-based electrodes reported in the literature and reflect on their developments.

4.6.1 Carbon-Based g-C_3N_4 Nanocomposites in Supercapacitors

The g-C_3N_4-based electrodes limit the practical usage when employed as SC electrodes, due to high charge transfer resistance [95].

However, the physiochemical properties of g-C_3N_4 nanocomposites with carbon-based materials can be significantly tuned by the adopted synthesis, morphology, functionalization, and introduction of surface defects to suit the application [97]. Such modifications are expected to overcome the shortfall of the g-C_3N_4 electrode. Another major factor influencing the activity of g-C_3N_4-based supercapacitors is the susceptibility to structural instability of g-C_3N_4 following charge–discharge. Many research groups attempted to graft onto g-C_3N_4, a conductive matrix of carbon such as graphene or rGO, which could also help in retaining the mechanical and structural stability of g-C_3N_4 electrodes. Chen et al. designed a symmetric SC with g-C_3N_4 and 3D interconnected networks of functionalized graphene (g-C_3N_4@G), which was synthesized by a hydrothermal method [98]. Supercapacitor device fabricated with two g-C_3N_4@G electrodes in 0.1 M of $LiClO_4$ delivered an energy density of 30 Wh kg^{-1} and a power density of 4.0 kW kg^{-1} at a current density of 4 A g^{-1} operating in a potential window of 1 V. The excellent performance of which was attributed to better charge transport in the g-C_3N_4@G electrode. Similarly, Lin et al. also conveyed a high specific capacitance for oxidized 3D g-C_3N_4 with functionalized graphene, owing to a greater number of oxygen and nitrogen-doped carbon sites [99]. The reduced graphene oxide acts as a skeleton for oxidized g-C_3N_4, resulting in high electron conductivity. Considering the mechanical and structural stability of g-C_3N_4 electrodes, the use of g-C_3N_4 nanocomposite flexible electrodes in supercapacitors was also demonstrated. The freestanding film of nano-fibrillated cellulose (NFC) could achieve a high specific capacitance of 51 F g^{-1} at 1 mA cm^{-2} [100]. Figure 4.7 (a) shows a schematic design for the preparation of NFC/g-C_3N_4 composite film via vacuum-induced self-assembly. The electrochemical characteristics of the fabricated freestanding NFC/g-C_3N_4 composite film electrodes with different g-C_3N_4 content are also shown in Fig. 4.7 (b–d). Hydrogen bonding and electrostatic bonds between g-C_3N_4 strands achieved the stability of the film.

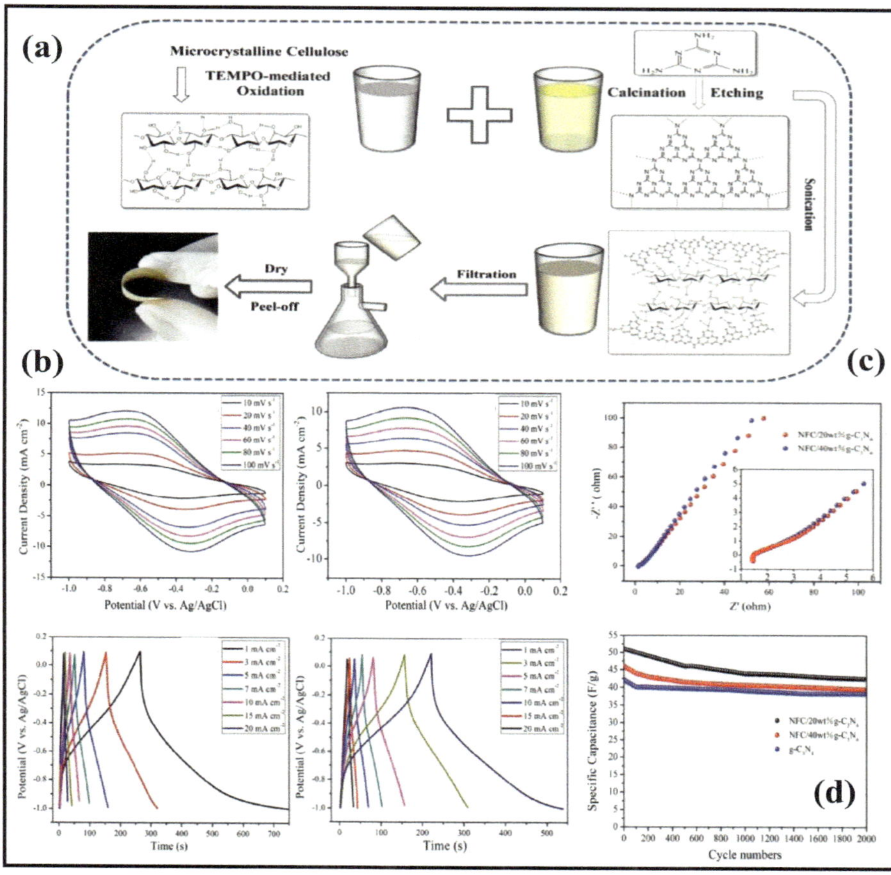

Figure 4.7 (a) Schematic diagram showing the production of NFC/g-C_3N_4 composite film based on NFC and g-C_3N_4 by vacuum-induced. (b) Electrochemical output of self-contained NFC/g-C_3N_4 composite film electrodes manufactured with a different g-C_3N_4 material, and CV curves at different scan rates with different g-C_3N_4 content and galvanostatic required to charge-discharge curves at different CDs with different g-C_3N_4 content. (c) EIS with NFC/g-C_3N_4 composite electrode with a different g-C_3N_4 material. (d) Period stability of NFC/g-C_3N_4 composite electrodes with a CD of 1 mA/cm^2. (Adapted from Ref. [100], Copyright 2019, *Vacuum*, Elsevier).

However, the EDLC nature of nanocomposite graphene with g-C_3N_4 offers low specific capacity, so there is a need to develop nanocomposites with higher specific capacity. The charge storage is low for the carbon composites, due to the surface charge separation process of the charge storage [93, 99, 101]. The integration of pseudocapacitive materials into g-C_3N_4 has been marked as an effective strategy to improve specific capacity, due to their high electrochemical active surface area.

4.6.2 g-C_3N_4 and Polymer Nanocomposites for Supercapacitors

Conductive polymers, such as PANI, which possess the advantages of being lightweight, low cost, flexibility, ease of synthesis, and tunable electrical conductivity, can be effective electrode materials for SCs. But, the applications of the same are limited, due to their low power density and poor cycling stability [102]. Combining g-C_3N_4 with conductive polymers to result in efficient charge transfer and prevent restacking of the layers of g-C_3N_4 (to expose active sites) provides an efficient strategy to improve the charge transfer and cycling stability. Zhou et al. attempted this by synthesizing polyaniline (PANI)/g-C_3N_4 composite [102]. Polyaniline was linked together into the g-C_3N_4 nanosheets acting as spacers to create gaps between consecutive g-C_3N_4 sheets. The PANI/g-C_3N_4 electrode revealed a specific capacitance of 584 F g^{-1} and energy density of ~81 Wh kg^{-1} with power density to the tune of 500 W kg^{-1}. The high cycling stability of the composite is credited to the structural integrity of PANI holding against the movement of ions into the nanosheets. The morphology or structure of the nanocomposite electrodes of conducting polymers with g-C_3N_4 plays quite crucial. The interconnected design both shortens the diffusion path lengths for the electrolyte ions and results in a rapid charge transfer reaction, thus enabling higher mass transfer efficiency and improving the limited cycle life of conventional SCs. While the pseudocapacitive charge storage of polymers increases specific capacity, the performance is limited to low energy density. In order to get higher specific capacity and high energy density in the SCs, battery-type electrodes have been developed and investigated.

4.6.3 g-C_3N_4 and Metal Oxides/Hydroxides Composites for Supercapacitors

Transition-metal oxides have been at the top list of battery-type electrode materials for SC applications, due to their high theoretical specific capacity. However, the transition metal oxides usually have relatively poor cycling stability that can be ascribed to the damage of morphology affected by the expansion and contraction of electrode materials during the charge/discharge processes. Combining layered metal hydroxides with conductive materials such as g-C_3N_4 and developing porous nanostructures are two main approaches to solving these problems. The nanocomposite of layered hydroxides and g-C_3N_4 can offset the structural stress from the electrodes. It also provides ion adsorption for charge transfer reactions and shortens the diffusion pathway of electrolyte ions. Metal oxides/hydroxides with g-C_3N_4 facilitate redox reactions, due to various oxidation states [101, 103]. The integration of metal oxides with g-C_3N_4 has been demonstrated to overcome large volume change, aggregation, and limited surface area deficiencies of metal oxide electrodes and simultaneously protect the g-C_3N_4 layers from restacking or expanding. For instance, RuO_2/g-C_3N_4@rGO aerogel composite synthesized via hydrothermal and lyophilization processes exhibited SC performance with a high specific capacitance of 704.3 F g^{-1} at a current density of 0.5 A g^{-1} [104]. The excellent electrochemical performance of RuO_2/g-C_3N_4@rGO has been assigned to the 3D cross-linked porous structure formed by RuO_2 and incorporated g-C_3N_4 layers into graphene sheets, which provided high surface area and more active sites than bulk g-C_3N_4 combined with graphene. Besides that, the high N-atom content of g-C_3N_4 provides plenty of electrochemically active sites. Low charge transfer resistance and high electrolyte diffusion have been observed for the composite. Similar to many other oxides, V_2O_5 nanospheres decorated on 3D porous g-C_3N_4 have provided the vertical charge transport pathway of a 3D porous carbon network, thus promoting electron diffusion during the rapid charge/discharge process [105].

The introduction of the g-C_3N_4 into hydroxide nanosheets can significantly enhance the capacitive performance and restrict the hydroxide nanosheets aggregation. Owing to this unique combination and structure, the g-C_3N_4-rGO/Ni$(OH)_2$ composite holds a large surface area with a suitable pore-size distribution, effectively accommodating the electrolyte ions migration and accelerating efficient electron transport [106]. With these impressive features, when used as an electrode for SC, the hybrid material exhibits high SC performance. Based on the feasibility of hydroxides, Adhikari et al. have reported the SC performance of g-C_3N_4/CuAl-LDH synthesized via a two-step process, i.e., electrodeposition and hydrothermal process [107]. The specific capacitances of the Ni-foam@Cu-Al LDH/g-C_3N_4 nanocomposite measured from the CV curve (770.98 F g^{-1} at 50 mV s^{-1}) and the galvanostatic charge/discharge curve (831.871 at 0.4 A g^{-1}) were significantly higher comparative pristine g-C_3N_4 or LDH. Moreover, the Ni-foam@Cu–Al LDH/g-C_3N_4 nanocomposite revealed a remarkable high-current capacitive behavior, and the capacitance retention could be maintained at 92.71% even after 5000 cycles of CV. When such nanocomposites having a 3D porous nanostructure were used as an electrode material for supercapacitors, they offered enhanced permeation of the electrolyte and allowed fast ion diffusion through the open and interconnected pore channels, which enhanced the super-capacitive performance of the materials such as enhanced a specific capacitance, high-rate capability, and excellent cycling stability. Thus, this research indicates that the 3D porous nanocomposite of g-C_3N_4 and Cu–Al LDH, prepared from a simple and convenient method, can be a competitive electrode material for the future development of supercapacitors. No doubt, the LDH/g-C_3N_4 composite supports the structural and chemical stability of electrodes during electrochemical reactions but it suffers from poor electronic conductivity.

4.6.4 g-C_3N_4 and Metal Chalcogenide Composites for Supercapacitors

Metal chalcogenides with g-C_3N_4 have a higher conductivity relative to their metal oxide/hydroxide equivalents. The combination of

2D metal chalcogenides with layered g-C_3N_4 has been shown to be a feasible technique to resolve the shortcomings of low electronic conductivity. MoS_2 is graphene-like with a layered structure that permits electrolyte ion insertion between its layers but the electrochemical performance of MoS_2 is not satisfactory, due to its limited electrical conductivity and structural instability during electrochemical reactions. Recently, the g-C_3N_4 nanosheets have been used as nanocomposite components with MoS_2 to increase surface area, wettability, and chemical affinity. MoS_2 offers plenty of active sites and exposure to electrode/electrolyte interaction, due to the high surface area. Combined with this, the porous nature of g-C_3N_4, acting as nanochannels on the electrodes, dramatically reduces the diffusion path and promotes high electron/ion transport developed metallic MoS_2 on porous g-C_3N_4 nanosheets using a simple calcination process [108]. A symmetric cell made of g-C_3N_4/MoS_2 composite was used to investigate the electrochemical performance of the electrode materials. The cell exhibited excellent performance with a maximum specific capacity of 198 C g^{-1} at a current density of 1 A g^{-1} and a specific energy density of 147 Wh kg^{-1} with a power density of 2691 W kg^{-1}. Besides, the cell exhibits extended cyclic stability with 89% capacitive retention after 8000 cycles. Thus, the efforts in developing electrode g-C_3N_4 nanocomposites with other layered materials are invaluable for SC applications.

The high surface area (220.7 m^2 g^{-1}) was exhibited by $NiCo_2S_4$ nanoparticles synthesized through a simple solvothermal process as a composite with self-repairing porous g-C_3N_4 nanosheets [109]. The fabricated hybrid composite electrode delivered excellent SC performance. The g-C_3N_4 acts as a conductivity bridge between high-capacitance (1557 F g^{-1} at a current density of 1 A g^{-1}) and long cycling stability (i.e., 7.4% loss after 10,000 cycles) nanoparticles. Figure 4.8 represents the synthesis of electrode materials and their electrochemical performance. Attractive energy storage density with ample capacity, stable operation, long cyclic life, and high-power density are generated by the g-C_3N_4 nanocomposite SCs. Thus, g-C_3N_4 is one of the potential candidates for energy storage application, and the steps taken by researchers to overcome the shortcomings of g-C_3N_4 are successful. Further research remains to be explored with regard to heteroatom doping in g-C_3N_4 sheets and their application in SCs. The development of g-C_3N_4-based nanocomposites can prove to be efficient and may be key to the commercialization of SC in the energy storage market.

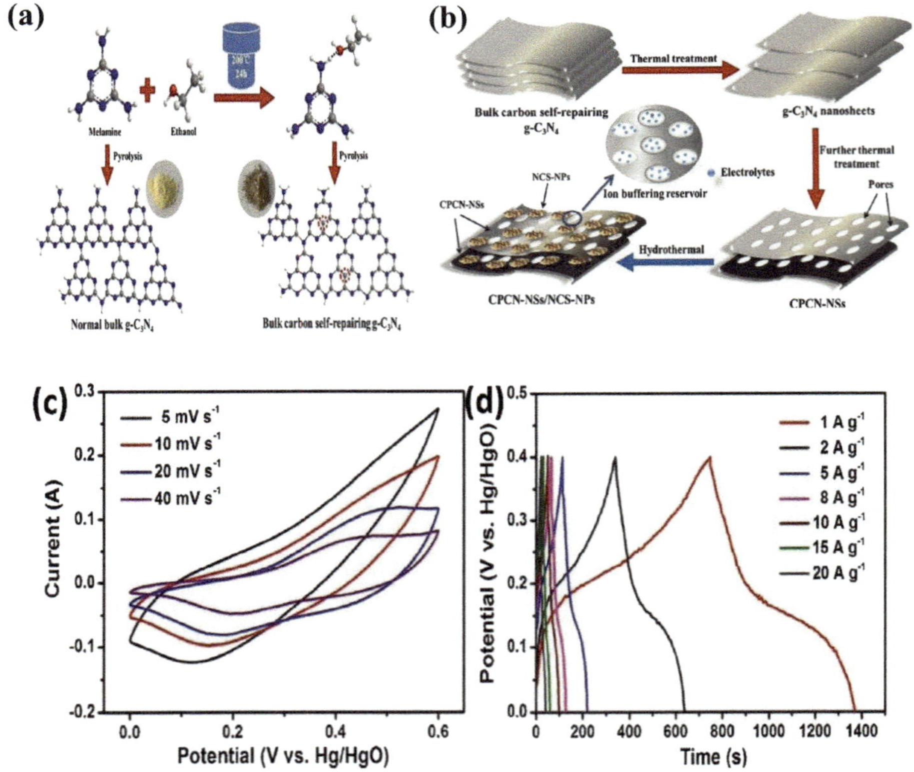

Figure 4.8 (a) Schematic representation of standard bulk g-C_3N_4 and bulk carbon self-repairing g-C_3N_4. (b) Schematic representation of carbon self-repairing porous g-C_3N_4 nanosheets/$NiCo_2S_4$ hybrid composite nanoparticles. (c) CV curves of hybrid electrodes at different scan rates. (d) Hybrid CD curves at different CDs. (Adapted from Ref. [109], Copyright 2017, *Electrochim.* Acta, Elsevier).

4.7 Summary and Perspective

A review of various g-C_3N_4-based materials used as electrodes for energy storage systems has been attempted in this chapter. With significant performance exhibited by the use of interlayer structures, the graphene-like structure of g-C_3N_4 has been more prevalent, due to its high thermal, chemical stability, and optical and photoelectric properties. Electrochemical applications of composites of g-C_3N_4 with various carbon-based materials, metal oxides/hydroxides, and chalcogenides have been discussed. From our investigations, we believe that the high electronic conductivity and good electrode binding provided by MXenes will mark better performance when employing MXenes with g-C_3N_4 interlayers. It is also imperative to study the electrode/electrolyte interlayer electrochemical performance, which will allow for a deeper understanding of the electrochemical mechanisms. Overall, electrodes based on g-C_3N_4 can effectively increase the electrochemical performance, resulting in advanced batteries and supercapacitors. For this reason, aspects that warrant further research on the effective utilization of 2D g-C_3N_4-based energy storage systems need to be researched. In the wake of low resource utilization, it is also imperative to inculcate theoretical calculations while searching for better electrode materials. Also, an extension of research to understand the underlying mechanism of the working of g-C_3N_4 in different environments will benefit in better design of composites. Ideally, restricting research to the 'trial and error' methods and mere reporting of performance may not be beneficial in extracting the full potential of such candidate materials.

References

1. L. Schlapbach, A. Züttel, Mobile applications, *Nature.* 414 (2001) 353–358. doi:10.1038/35104634.
2. K. Hari Prasad, S. Subramanian, T.N. Sairam, G. Amarendra, E.S. Srinadhu, N. Satyanarayana, Structural, electrical and dielectric properties of nanocrystalline $LiMgBO_3$ particles synthesized by Pechini process, *J. Alloys Compd.* 718 (2017) 459–470. doi:10.1016/j.jallcom.2017.05.157.
3. K. Hari Prasad, S. Vinoth, P. Jena, M. Venkateswarlu, N. Satyanarayana, Structural characterization and impedance studies of PbO nanofibers synthesized by electrospinning technique, *Mater. Chem. Phys.* 194 (2017) 188–197. doi:10.1016/j.matchemphys.2017.03.040.

4. M. Liang, L. Zhi, Graphene-based electrode materials for rechargeable lithium batteries, *J. Mater. Chem.* 19 (2009) 5871–5878. doi:10.1039/b901551e.

5. R. Zhao, Z. Qian, Z. Liu, D. Zhao, X. Hui, G. Jiang, C. Wang, L. Yin, Molecular-level heterostructures assembled from layered black phosphorene and Ti_3C_2 MXene as superior anodes for high-performance sodium ion batteries, *Nano Energy.* 65 (2019). doi:10.1016/j.nanoen.2019.104037.

6. D. Han, J. Zhang, Z. Weng, D. Kong, Y. Tao, F. Ding, D. Ruan, Q.H. Yang, Two-dimensional materials for lithium/sodium-ion capacitors, *Mater. Today Energy.* 11 (2019) 30–45. doi:10.1016/j.mtener.2018.10.013.

7. J. Theerthagiri, G. Durai, K. Karuppasamy, P. Arunachalam, V. Elakkiya, P. Kuppusami, T. Maiyalagan, H.S. Kim, Recent advances in 2-D nanostructured metal nitrides, carbides, and phosphides electrodes for electrochemical supercapacitors – A brief review, *J. Ind. Eng. Chem.* 67 (2018) 12–27. doi:10.1016/j.jiec.2018.06.038.

8. G. Liu, T. Wang, H. Zhang, X. Meng, D. Hao, K. Chang, P. Li, T. Kako, J. Ye, Nature-inspired environmental "phosphorylation" boosts photocatalytic H_2 production over carbon nitride nanosheets under visible-light irradiation, *Angew. Chemie - Int. Ed.* 54 (2015) 13561–13565. doi:10.1002/anie.201505802.

9. A.Y. Liu, M.L. Cohen, Prediction of new low compressibility solids, *Science (80-.).* 245 (1989) 841–842. doi:10.1126/science.245.4920.841.

10. F. Goettmann, A. Fischer, M. Antonietti, A. Thomas, Metal-free catalysis of sustainable Friedel-Crafts reactions: Direct activation of benzene by carbon nitrides to avoid the use of metal chlorides and halogenated compounds, *Chem. Commun.* (2006) 4530–4532. doi:10.1039/b608532f.

11. D.M. Teter, R.J. Hemley, Low-compressibility carbon nitrides, *Science (80-.).* 271 (1996) 53–55. doi:10.1126/science.271.5245.53.

12. J. Liu, W. Li, L. Duan, X. Li, L. Ji, Z. Geng, K. Huang, L. Lu, L. Zhou, Z. Liu, W. Chen, L. Liu, S. Feng, Y. Zhang, A graphene-like oxygenated carbon nitride material for improved cycle-life lithium/sulfur batteries, *Nano Lett.* 15 (2015) 5137–5142. doi:10.1021/acs.nanolett.5b01919.

13. J. Zhang, N. Zhao, M. Zhang, Y. Li, P.K. Chu, X. Guo, Z. Di, X. Wang, H. Li, Flexible and ion-conducting membrane electrolytes for solid-state lithium batteries: Dispersion of garnet nanoparticles in insulating polyethylene oxide, *Nano Energy.* 28 (2016) 447–454. doi:10.1016/j.nanoen.2016.09.002.

14. L. Shi, T. Wang, H. Zhang, K. Chang, J. Ye, Electrostatic self-assembly of nanosized carbon nitride nanosheet onto a zirconium metal-organic framework for enhanced photocatalytic CO_2 reduction, *Adv. Funct. Mater.* 25 (2015) 5360–5367. doi:10.1002/adfm.201502253.

15. J. Fang, H. Fan, M. Li, C. Long, Nitrogen self-doped graphitic carbon nitride as efficient visible light photocatalyst for hydrogen evolution, *J. Mater. Chem. A.* 3 (2015) 13819–13826. doi:10.1039/c5ta02257f.

16. T.Y. Ma, Y. Tang, S. Dai, S.Z. Qiao, Proton-functionalized two-dimensional graphitic carbon nitride nanosheet: An excellent metal-/label-free biosensing platform, *Small.* 10 (2014) 2382–2389. doi:10.1002/smll.201303827.

17. J. Zhu, Y. Wei, W. Chen, Z. Zhao, A. Thomas, Graphitic carbon nitride as a metal-free catalyst for NO decomposition, *Chem. Commun.* 46 (2010) 6965–6967. doi:10.1039/c0cc01432j.

18. A. Thomas, A. Fischer, F. Goettmann, M. Antonietti, J.O. Müller, R. Schlögl, J.M. Carlsson, Graphitic carbon nitride materials: Variation of structure and morphology and their use as metal-free catalysts, *J. Mater. Chem.* 18 (2008) 4893–4908. doi:10.1039/b800274f.

19. Y. Luo, Y. Yan, S. Zheng, H. Xue, H. Pang, Graphitic carbon nitride based materials for electrochemical energy storage. (2019) 901–924. doi:10.1039/c8ta08464e.

20. D.P. Dubal, O. Ayyad, V. Ruiz, P. Gómez-Romero, Hybrid energy storage: The merging of battery and supercapacitor chemistries, *Chem. Soc. Rev.* 44 (2015) 1777–1790. doi:10.1039/c4cs00266k.

21. R.M. Humana, M.G. Ortiz, J.E. Thomas, S.G. Real, M. Sedlarikova, J. Vondrak, A. Visintin, Preparation and characterization of graphite anode for lithium ion batteries, *ECS Trans.* 63 (2014) 91–97. doi:10.1149/06301.0091ecst.

22. J.-M. Tarascon, M. Armandt, Issues and challenges facing rechargeable lithium batteries, *Nat. Insight.* 414 (2001) 359–367.

23. H. Pan, *Graphitic Carbon Nitride Nanotubes As Li-Ion Battery Materials: A First-Principles Study.* (2014).

24. G.M. Veith, L.A. Adamczyk, B. Guo, S.S. Brown, X. Sun, A.A. Albert, J.R. Humble, C.E. Barnes, M.J. Bojdys, S. Dai, N.J. Dudney, *Electrochemical and Solid-State Lithiation of Graphitic C_3N_4.* (2013) 3–8. doi:10.1021/cm303870x.

25. M. Hankel, D. Ye, L. Wang, D.J. Searles, *Lithium and Sodium Storage on Graphitic Carbon Nitride.* (2015) 4–10. doi:10.1021/acs.jpcc.5b07572.

26. J. Chen, Z. Mao, L. Zhang, D. Wang, R. Xu, L. Bie, B.D. Fahlman, *Nitrogen-Deficient Graphitic Carbon Nitride with Enhanced Performance for Lithium Ion Battery Anodes.* (2017). doi:10.1021/acsnano.7b07116.

27. M. Wu, Q. Wang, Q. Sun, P. Jena, *Functionalized Graphitic Carbon Nitride for Efficient Energy Storage.* (2013). doi:10.1021/jp311972f.

28. L. Qie, W. Chen, Z. Wang, Q. Shao, X. Li, L. Yuan, Nitrogen-doped porous carbon nanofiber webs as anodes for lithium ion batteries with a superhigh capacity and rate capability. (2012) 2047–2050. doi:10.1002/adma.201104634.

29. H. Tao, L. Xiong, S. Du, Y. Zhang, X. Yang, Interwoven N and P dual-doped hollow carbon fibers/graphitic carbon nitride: An ultrahigh capacity and rate anode for Li and Na ion batteries, *Carbon N. Y.* 122 (2017) 54–63. doi:10.1016/j.carbon.2017.06.040.

30. Y. Fu, J. Zhu, C. Hu, X. Wu, X. Wang, Nanoscale covalently coupled hybrid of graphitic carbon nitride with reduced graphene oxide as a superior. (2014) 12555–12564. doi:10.1039/c4nr03145h.

31. W. Xu, J. Wang, F. Ding, X. Chen, E. Nasybulin, Y. Zhang, J.G. Zhang, Lithium metal anodes for rechargeable batteries, *Energy Environ. Sci.* 7 (2014) 513–537. doi:10.1039/c3ee40795k.

32. M. Shi, T. Wu, X. Song, J. Liu, L. Zhao, P. Zhang, Active Fe_2O_3 nanoparticles encapsulated in porous g-C_3N_4/graphene sandwich-type nanosheets as a superior anode for high-performance lithium-ion batteries. (2016) 10666–10672. doi:10.1039/c6ta03533g.

33. B. Joshi, E. Samuel, T. Kim, C. Park, Y. Kim, M.T. Swihart, W. Young, S.S. Yoon, Supersonically spray-coated zinc ferrite/graphitic-carbon nitride composite as a stable high-capacity anode material for lithium-ion batteries, *J. Alloys Compd.* 768 (2018) 525–534. doi:10.1016/j.jallcom.2018.07.027.

34. D. Versaci, J. Amici, C. Francia, S. Bodoardo, Simple approach using g-C_3N_4 to enable SnO_2 anode high rate performance for Li ion battery, *Solid State Ionics.* 346 (2020) 115210. doi:10.1016/j.ssi.2019.115210.

35. G. Wang, Z. Wen, Y.E. Yang, J. Yin, W. Kong, S. Li, J. Sun, S. Ji, Ultra-long life Si@rGO/g-C_3N_4 with a multiply synergetic effect as an anode material for lithium-ion batteries, *J. Mater. Chem. A.* 6 (2018) 7557–7565. doi:10.1039/c8ta00539g.

36. C. Senthil, S.C. Kishore, M. Sasidharan, Applied surface science ultrathin MoS_2 sheets supported on N-rich carbon nitride nanospheres with enhanced lithium storage properties, *Appl. Surf. Sci.* 410 (2017) 215–224. doi:10.1016/j.apsusc.2017.03.102.

37. Y. Hou, Z. Wen, S. Cui, X. Guo, J. Chen, Constructing 2D porous graphitic C_3N_4 nanosheets/nitrogen-doped graphene/layered MoS_2 ternary nanojunction with enhanced photoelectrochemical activity. (2013) 6291–6297. doi:10.1002/adma.201303116.

38. L. Yin, R. Cheng, Q. Song, J. Yang, X. Kong, J. Huang, Y. Lin, H. Ouyang, Construction of nanoflower SnS_2 anchored on g-C_3N_4 nanosheets composite as highly efficient anode for lithium ion batteries, *Electrochim. Acta.* 293 (2019) 408–418. doi:10.1016/j.electacta.2018.10.020.

39. S. Arif, S. Shah, A.R. Park, A. Rauf, S.H. Hong, Y. Choi, J. Park, J. Kim, W. Kim, P.J. Yoo, Highly interdigitated and porous architected ion battery anodes. (2017) 3125–3135. doi:10.1039/c6ra25886g.

40. Y. Wang, X. Ding, L. Huang, Y. Xu, Y. Zuo, S. Kang, L. Cui, Critical influence of carbon nitride self-assembly coating on the electrochemical performance of SnO_2-TiO_2 nanocomposite anode material for lithium ion battery, 11 (2016) 2461–2472.

41. V. Palomares, P. Serras, I. Villaluenga, K.B. Hueso, J. Carretero-González, T. Rojo, Na-ion batteries, recent advances and present challenges to become low cost energy storage systems, *Energy Environ. Sci.* 5 (2012) 5884–5901. doi:10.1039/c2ee02781j.

42. G.M. Weng, Y. Xie, H. Wang, C. Karpovich, J. Lipton, J. Zhu, J. Kong, L.D. Pfefferle, A.D. Taylor, A promising carbon/g-C_3N_4 composite negative electrode for a long-life sodium-ion battery, *Angew. Chemie - Int. Ed.* 58 (2019) 13727–13733. doi:10.1002/anie.201905803.

43. M.Y. Wang, J. Li, A first-principles study of lithium and sodium storage in two-dimensional graphitic carbon nitride, *Xinxing Tan Cailiao/New Carbon Mater.* 33 (2018) 510–515. doi:10.1016/S1872-5805(18)60353-1.

44. G.M. Weng, Y. Xie, H. Wang, C. Karpovich, J. Lipton, J. Zhu, J. Kong, L.D. Pfefferle, A.D. Taylor, A promising carbon/g-C_3N_4 composite negative electrode for a long-life sodium-ion battery, *Angew. Chemie - Int. Ed.* 58 (2019) 13727–13733. doi:10.1002/anie.201905803.

45. M. Molaei, S.M. Mousavi-Khoshdel, M. Ghiasi, Exploring the effect of phosphorus doping on the utility of g-C_3N_4 as an electrode material in Na-ion batteries using DFT method, *J. Mol. Model.* 25 (2019) 1–8. doi:10.1007/s00894-019-4109-1.

46. S. Wang, Y. Zhu, M. Jiang, J. Cui, Y. Zhang, W. He, Interconnected $Na_2Ti_3O_7$ nanotube/g-C_3N_4/graphene network as high performance anode materials for sodium storage, *Int. J. Hydrogen Energy.* 45 (2020) 19611–19619. doi:10.1016/j.ijhydene.2020.05.133.

47. Y. Zhou, S. Zhang, J. Xu, Y. Zhang, Construction of MoS_2-nitrogen-deficient graphitic carbon nitride anode toward high performance sodium-ions batteries, *Mater. Lett.* 273 (2020) 127890. doi:10.1016/j.matlet.2020.127890.

48. B. Li, B. Xi, Z. Feng, Y. Lin, J. Liu, J. Feng, Y. Qian, S. Xiong, Hierarchical porous nanosheets constructed by graphene-coated, interconnected TiO_2 nanoparticles for ultrafast sodium storage, *Adv. Mater.* 30 (2018) 1–9. doi:10.1002/adma.201705788.

49. Y. Luo, Y. Yan, S. Zheng, H. Xue, H. Pang, Graphitic carbon nitride based materials for electrochemical energy storage, *J. Mater. Chem. A.* 7 (2019) 901–924. doi:10.1039/c8ta08464e.

50. X. Wang, X. Chen, A. Thomas, X. Fu, M. Antonietti, Metal-containing carbon nitride compounds: a new functional organic-metal hybrid material, *Adv. Mater.* 21 (2009) 1609–1612. doi:10.1002/adma.200802627.

51. S.M. Lyth, Y. Nabae, N.M. Islam, S. Kuroki, M. Kakimoto, J. Ozaki, S. Miyata, Electrochemical oxygen reduction on carbon nitride. 28 (2010) 11–26. doi:10.1149/1.3502333.

52. Y. Zheng, J. Liu, J. Liang, M. Jaroniec, S.Z. Qiao, Graphitic carbon nitride materials: Controllable synthesis and applications in fuel cells and photocatalysis, *Energy Environ. Sci.* 5 (2012) 6717–6731. doi:10.1039/c2ee03479d.

53. Y. Zheng, Y. Jiao, J. Chen, J. Liu, J. Liang, A. Du, W. Zhang, Z. Zhu, S.C. Smith, M. Jaroniec, G.Q. Lu, S.Z. Qiao, Nanoporous graphitic-C_3N_4@carbon metal-free electrocatalysts for highly efficient oxygen reduction, *J. Am. Chem. Soc.* 133 (2011) 20116–20119. doi:10.1021/ja209206c.

54. X. Chen, R. Hu, DFT-based study of single transition metal atom doped g-C_3N_4 as alternative oxygen reduction reaction catalysts, *Int. J. Hydrogen Energy.* 44 (2019) 15409–15416. doi:10.1016/j.ijhydene.2019.04.057.

55. Q. Liu, J. Zhang, Graphene supported Co-g-C_3N_4 as a novel metal-macrocyclic electrocatalyst for the oxygen reduction reaction in fuel cells, *Langmuir.* 29 (2013) 3821–3828. doi:10.1021/la400003h.

56. S. Sarkar, S.S. Sumukh, K. Roy, N. Kamboj, T. Purkait, M. Das, R.S. Dey, Facile one step synthesis of Cu-g-C_3N_4 electrocatalyst realized oxygen reduction reaction with excellent methanol crossover impact and durability, *J. Colloid Interface Sci.* 558 (2019) 182–189. doi:10.1016/j.jcis.2019.09.107.

57. N. Bhandary, P.P. Ingole, S. Basu, Facile solid-state synthesis of Ag/g-C_3N_4 reinforced graphene oxide for enhanced electrocatalysis of oxygen reduction reaction in alkaline fuel cell, *ChemistrySelect.* 2 (2017) 8151–8157. doi:10.1002/slct.201700926.

58. S. Soren, I. Hota, A.K. Debnath, D.K. Aswal, K.S.K. Varadwaj, P. Parhi, Oxygen reduction Reaction activity of microwave mediated solvothermal synthesized CeO_2/g-C_3N_4 nanocomposite, *Front. Chem.* 7 (2019) 1–10. doi:10.3389/fchem.2019.00403.

59. J. Tian, Q. Liu, A.M. Asiri, K.A. Alamry, X. Sun, Ultrathin graphitic C_3N_4 nanosheets/graphene composites: Efficient organic electrocatalyst for oxygen evolution reaction, *ChemSusChem.* 7 (2014) 2125–2130. doi:10.1002/cssc.201402118.

60. M. Balaish, A. Kraytsberg, Y. Ein-Eli, A critical review on lithium-air battery electrolytes, *Phys. Chem. Chem. Phys.* 16 (2014) 2801–2822. doi:10.1039/c3cp54165g.

61. W. Bin Luo, S.L. Chou, J.Z. Wang, Y.C. Zhai, H.K. Liu, A metal-free, free-standing, macroporous graphene@g-C_3N_4 composite air electrode for high-energy lithium oxygen batteries, *Small.* 11 (2015) 2817–2824. doi:10.1002/smll.201403535.

62. J. Yi, K. Liao, C. Zhang, T. Zhang, F. Li, H. Zhou, Facile in situ preparation of graphitic-C_3N_4@carbon paper as an efficient metal-free cathode for nonaqueous Li-O_2 battery, *ACS Appl. Mater. Interfaces.* 7 (2015) 10823–10827. doi:10.1021/acsami.5b01727.

63. S. Kumar, A. Jena, Y.C. Hu, C. Liang, W. Zhou, T.F. Hung, W.S. Chang, H. Chang, R.S. Liu, Cobalt diselenide nanorods grafted on graphitic carbon nitride: A synergistic catalyst for oxygen reactions in rechargeable Li-O_2 batteries, *ChemElectroChem.* 5 (2018) 29–35. doi:10.1002/celc.201700909.

64. Y. Hang, C. Zhang, X. Luo, Y. Xie, S. Xin, Y. Li, D. Zhang, J.B. Goodenough, A-MnO_2 nanorods supported on porous graphitic carbon nitride as efficient electrocatalysts for lithium-air batteries, *J. Power Sources.* 392 (2018) 15–22. doi:10.1016/j.jpowsour.2018.04.078.

65. Y. Wu, T. Wang, Y. Zhang, S. Xin, X. He, D. Zhang, J. Shui, Electrocatalytic performances of g-C_3N_4-$LaNiO_3$ composite as bi-functional catalysts for lithium-oxygen batteries, *Sci. Rep.* 6 (2016) 4–11. doi:10.1038/srep24314.

66. P. Lou, Z. Cui, X. Guo, Achieving highly stable Li-O_2 battery operation by designing a carbon nitride-based cathode towards a stable reaction interface, *J. Mater. Chem. A.* 5 (2017) 18207–18213. doi:10.1039/c7ta05009g.

67. J.B. Park, S.H. Lee, H.G. Jung, D. Aurbach, Y.K. Sun, Redox mediators for Li-O_2 batteries: Status and perspectives, *Adv. Mater.* 30 (2018) 1–13. doi:10.1002/adma.201704162.

68. H.D. Lim, H. Song, J. Kim, H. Gwon, Y. Bae, K.Y. Park, J. Hong, H. Kim, T. Kim, Y.H. Kim, X. Leprõ, R. Ovalle-Robles, R.H. Baughman, K. Kang, Superior rechargeability and efficiency of lithium-oxygen batteries: Hierarchical air electrode architecture combined with a soluble catalyst, *Angew. Chemie - Int. Ed.* 53 (2014) 3926–3931. doi:10.1002/anie.201400711.

69. Y. Liu, N. Li, S. Wu, K. Liao, K. Zhu, J. Yi, H. Zhou, Reducing the charging voltage of a Li-O_2 battery to 1.9 v by incorporating a photocatalyst, *Energy Environ. Sci.* 8 (2015) 2664–2667. doi:10.1039/c5ee01958c.

70. J. Fu, R. Liang, G. Liu, A. Yu, Z. Bai, L. Yang, Z. Chen, Recent progress in electrically rechargeable zinc–air batteries, *Adv. Mater.* 31 (2019) 1–13. doi:10.1002/adma.201805230.

71. J.E. Park, M.J. Kim, M.S. Lim, S.Y. Kang, J.K. Kim, S.H. Oh, M. Her, Y.H. Cho, Y.E. Sung, Graphitic carbon nitride-carbon nanofiber as oxygen catalyst in anion-exchange membrane water electrolyzer and rechargeable metal–air cells, *Appl. Catal. B Environ.* 237 (2018) 140–148. doi:10.1016/j.apcatb.2018.05.073.

72. S.S. Shinde, C.H. Lee, A. Sami, D.H. Kim, S.U. Lee, J.H. Lee, Scalable 3-D carbon nitride sponge as an efficient metal-free bifunctional oxygen electrocatalyst for rechargeable Zn-Air batteries, *ACS Nano.* 11 (2017) 347–357. doi:10.1021/acsnano.6b05914.

73. S.S. Shinde, J.Y. Yu, J.W. Song, Y.H. Nam, D.H. Kim, J.H. Lee, Highly active and durable carbon nitride fibers as metal-free bifunctional oxygen electrodes for flexible Zn-air batteries, *Nanoscale Horizons.* 2 (2017) 333–341. doi:10.1039/c7nh00058h.

74. W. Niu, Z. Li, K. Marcus, L. Zhou, Y. Li, R. Ye, K. Liang, Y. Yang, Surface-modified porous carbon nitride composites as highly efficient electrocatalyst for Zn-Air batteries, *Adv. Energy Mater.* 8 (2018) 1–8. doi:10.1002/aenm.201701642.

75. W. Niu, K. Marcus, L. Zhou, Z. Li, L. Shi, K. Liang, Y. Yang, Enhancing electron transfer and electrocatalytic activity on crystalline carbon-conjugated g-C_3N_4, *ACS Catal.* 8 (2018) 1926–1931. doi:10.1021/acscatal.8b00026.

76. Z. Li, L. Yuan, Z. Yi, Y. Sun, Y. Liu, Y. Jiang, Y. Shen, Y. Xin, Z. Zhang, Y. Huang, Insight into the electrode mechanism in lithium-sulfur batteries with ordered microporous carbon confined sulfur as the cathode, *Adv. Energy Mater.* 4 (2014) 1–8. doi:10.1002/aenm.201301473.

77. C.S. Smaller, Z. Li, Y. Jiang, L. Yuan, Z. Yi, C. Wu, Y. Liu, P. Strasser, Y. Huang, A highly ordered meso @ microporous sulfur core à shell structured, *ACS Nano.* 8 (2014) 9295–9303. http://pubs.acs.org/doi/pdfplus/10.1021/nn503220h.

78. X. Ji, K.T. Lee, L.F. Nazar, A highly ordered nanostructured carbon-sulphur cathode for lithium-sulphur batteries, *Nat. Mater.* 8 (2009) 500–506. doi:10.1038/nmat2460.

79. C. Zhang, H. Bin Wu, C. Yuan, Z. Guo, X.W. Lou, Confining sulfur in double-shelled hollow carbon spheres for lithium-sulfur batteries, *Angew. Chemie - Int. Ed.* 51 (2012) 9592–9595. doi:10.1002/anie.201205292.

80. N. Jayaprakash, J. Shen, S.S. Moganty, A. Corona, L.A. Archer, Porous hollow carbon@sulfur composites for high-power lithium-sulfur batteries, *Angew. Chemie - Int. Ed.* 50 (2011) 5904–5908. doi:10.1002/anie.201100637.

81. G. He, S. Evers, X. Liang, M. Cuisinier, A. Garsuch, L. Nazar, Tailoring porosity in carbon, *ACS Nano.* 7 (2013) 10920–10930.

82. C. Wang, X. Wang, Y. Yang, A. Kushima, J. Chen, Y. Huang, J. Li, Slurryless Li_2S/reduced graphene oxide cathode paper for high-performance lithium sulfur battery, *Nano Lett.* 15 (2015) 1796–1802. doi:10.1021/acs.nanolett.5b00112.

83. H. Wang, Y. Yang, Y. Liang, J.T. Robinson, Y. Li, A. Jackson, Nl200658a. Pdf, *Nano Lett.* 11 (2011) 2644–2647.

84. Q. Pang, L.F. Nazar, Long-life and high-areal-capacity Li-S batteries enabled by a light-weight polar host with intrinsic polysulfide adsorption, *ACS Nano.* 10 (2016) 4111–4118. doi:10.1021/acsnano.5b07347.

85. Y. Gong, C. Fu, G. Zhang, H. Zhou, Y. Kuang, Three-dimensional porous C_3N_4 nanosheets@reduced graphene oxide network as sulfur hosts for high performance lithium-sulfur batteries, *Electrochim. Acta.* 256 (2017) 1–9. doi:10.1016/j.electacta.2017.10.032.

86. Y.S. Su, A. Manthiram, Lithium-sulphur batteries with a microporous carbon paper as a bifunctional interlayer, *Nat. Commun.* 3 (2012) 1166. doi:10.1038/ncomms2163.

87. C.Y. Fan, H.Y. Yuan, H.H. Li, H.F. Wang, W.L. Li, H.Z. Sun, X.L. Wu, J.P. Zhang, The effective design of a polysulfide-trapped separator at the molecular level for high energy density Li-S batteries, *ACS Appl. Mater. Interfaces.* 8 (2016) 16108–16115. doi:10.1021/acsami.6b04578.

88. L. Qu, P. Liu, Y. Yi, T. Wang, P. Yang, X. Tian, M. Li, B. Yang, S. Dai, Enhanced cycling performance for lithium–sulfur batteries by a laminated 2D g-C_3N_4/graphene cathode interlayer, *ChemSusChem.* 12 (2019) 213–223. doi:10.1002/cssc.201802449.

89. M. Chen, X. Zhao, Y. Li, P. Zeng, H. Liu, H. Yu, M. Wu, Z. Li, D. Shao, C. Miao, G. Chen, H. Shu, Y. Pei, X. Wang, Kinetically elevated redox conversion of polysulfides of lithium-sulfur battery using a separator modified with transition metals coordinated g-C_3N_4 with carbon-conjugated, *Chem. Eng. J.* 385 (2020) 123905. doi:10.1016/j.cej.2019.123905.

90. D.R. Deng, C.D. Bai, F. Xue, J. Lei, P. Xu, M. Sen Zheng, Q.F. Dong, Multifunctional ion-sieve constructed by 2D materials as an interlayer for Li-S batteries, *ACS Appl. Mater. Interfaces.* 11 (2019) 11474–11480. doi:10.1021/acsami.8b22660.

91. H.T. Das, K. Mahendraprabhu, T. Maiyalagan, P. Elumalai, Performance of solid-state hybrid energy-storage device using reduced graphene-oxide anchored sol-gel derived Ni/NiO nanocomposite, *Sci. Rep.* 7 (2017) 1–14. doi:10.1038/s41598-017-15444-z.

92. M.S. Uddin, H. Tanaya Das, T. Maiyalagan, P. Elumalai, Influence of designed electrode surfaces on double layer capacitance in aqueous electrolyte: Insights from standard models, *Appl. Surf. Sci.* 449 (2018) 445–453. doi:10.1016/j.apsusc.2017.12.088.

93. D. Wang, Y. Wang, Y. Chen, W. Liu, H. Wang, P. Zhao, Y. Li, J. Zhang, Y. Dong, S. Hu, J. Yang, Coal tar pitch derived N-doped porous carbon nanosheets by the in-situ formed g-C_3N_4 as a template for supercapacitor electrodes, *Electrochim. Acta.* 283 (2018) 132–140. doi:10.1016/j.electacta.2018.06.151.

94. M. Gupta, P.K. Singh, B. Bhattacharya, Y.M. Shulga, N.Y. Shulga, Y. Kumar, Progress, status and prospects of non-porous, heteroatom-doped carbons for supercapacitors and other electrochemical applications, *Appl. Phys. A Mater. Sci. Process.* 125 (2019) 1–15. doi:10.1007/s00339-019-2400-8.

95. Y. Ding, Y. Tang, L. Yang, Y. Zeng, J. Yuan, T. Liu, S. Zhang, C. Liu, S. Luo, Porous nitrogen-rich carbon materials from carbon self-repairing g-C_3N_4 assembled with graphene for high-performance supercapacitor, *J. Mater. Chem. A.* 4 (2016) 14307–14315. doi:10.1039/c6ta05267c.

96. H. Pan, Graphitic carbon nitride nanotubes as li-ion battery materials: A first-principles study, *J. Phys. Chem. C.* 118 (2014) 9318–9323. doi:10.1021/jp4122722.

97. M. Tahir, C. Cao, N. Mahmood, F.K. Butt, A. Mahmood, F. Idrees, S. Hussain, M. Tanveer, Z. Ali, I. Aslam, Multifunctional g-C_3N_4 nanofibers: A template-free fabrication and enhanced optical, electrochemical, and photocatalyst properties, *ACS Appl. Mater. Interfaces.* 6 (2014) 1258–1265. doi:10.1021/am405076b.

98. Q. Chen, Y. Zhao, X. Huang, N. Chen, L. Qu, Three-dimensional graphitic carbon nitride functionalized graphene-based high-performance supercapacitors, *J. Mater. Chem. A.* 3 (2015) 6761–6766. doi:10.1039/c5ta00734h.

99. R. Lin, Z. Li, D.I. Abou El Amaiem, B. Zhang, D.J.L. Brett, G. He, I.P. Parkin, A general method for boosting the supercapacitor performance of graphitic carbon nitride/graphene hybrids, *J. Mater. Chem. A.* 5 (2017) 25545–25554. doi:10.1039/c7ta09492b.

100. F. Li, X. Gu, K. Zhang, T.T. Nguyen, M. Guo, Fabrication of freestanding NFC/g-C_3N_4 composite film as supercapacitor electrode via vacuum-induced self-assembly, *Vacuum.* 160 (2019) 54–59. doi:10.1016/j.vacuum.2018.10.005.

101. C. Yang, J. Liu, X. Song, L. Zhao, P. Zhang, L. Gao, Pseudocapacitive quantum dots confined in sacrificial g-C_3N_4 derived carbon nanosheets for high performance ionic liquid-based supercapacitors, *Mater. Lett.* 266 (2020). doi:10.1016/j.matlet.2020.127498.

102. S.X. Zhou, X.Y. Tao, J. Ma, L.T. Guo, Y.B. Zhu, H.L. Fan, Z.S. Liu, X.Y. Wei, Synthesis of flower-like PANI/g-C_3N_4 nanocomposite as supercapacitor electrode, *Vacuum.* 149 (2018) 175–179. doi:10.1016/j.vacuum.2017.12.019.

103. S.V.P. Vattikuti, B.P. Reddy, C. Byon, J. Shim, Carbon/CuO nanosphere-anchored g-C_3N_4 nanosheets as ternary electrode material for supercapacitors, *J. Solid State Chem.* 262 (2018) 106–111. doi:10.1016/j.jssc.2018.03.019.

104. J. Zhang, J. Ding, C. Li, B. Li, D. Li, Z. Liu, Q. Cai, J. Zhang, Y. Liu, Fabrication of novel ternary three-dimensional RuO_2/graphitic-C_3N_4@ reduced graphene oxide aerogel composites for supercapacitors, *ACS Sustain. Chem. Eng.* 5 (2017) 4982–4991. doi:10.1021/acssuschemeng.7b00358.

105. Y. Zhou, L. Sun, D. Wu, X. Li, J. Li, P. Huo, H. Wang, Y. Yan, Preparation of 3D porous g-C_3N_4@V_2O_5 composite electrode via simple calcination and chemical precipitation for supercapacitors, *J. Alloys Compd.* 817 (2020). doi:10.1016/j.jallcom.2019.152707.

106. L. Li, J. Qin, H. Bi, S. Gai, F. He, P. Gao, Y. Dai, X. Zhang, D. Yang, P. Yang, $Ni(OH)_2$ nanosheets grown on porous hybrid g-C_3N_4/RGO network

as high performance supercapacitor electrode, *Sci. Rep.* 7 (2017). doi:10.1038/srep43413.

107. S.P. Adhikari, G.P. Awasthi, K.S. Kim, C.H. Park, C.S. Kim, Synthesis of three-dimensional mesoporous Cu-Al layered double hydroxide/g-C_3N_4 nanocomposites on Ni-foam for enhanced supercapacitors with excellent long-term cycling stability, *Dalt. Trans.* 47 (2018) 4455–4466. doi:10.1039/c7dt04192f.

108. N. Joseph, A.C. Bose, Metallic MoS_2 grown on porous g-C_3N_4 as an efficient electrode material for supercapattery application, *Electrochim. Acta.* 301 (2019) 401–410. doi:10.1016/j.electacta.2019.01.155.

109. W. Guo, J. Wang, C. Fan, Z. Chen, P. Liu, D. Zhu, Z. Xu, L. Pang, T. Li, Synthesis of carbon self-repairing porous g-C_3N_4 nanosheets/$NiCo_2S_4$ nanoparticles hybrid composite as high-performance electrode materials for supercapacitors, *Electrochim. Acta.* 253 (2017) 68–77. doi:10.1016/j.electacta.2017.09.025.

Chapter 5

Ecofriendly Cobalt-Free, Li- and Mn-Rich Layered Materials for High-Performance Lithium-Ion Batteries

Prettencia Leonard Joseph,[a] Soundarrajan Elumalai,[a] Kalaivani Raman,[a] and Raghu Subashchandrabose[b]

[a]*Department of Chemistry, Vels Institute of Science, Technology & Advanced Studies, Chennai, Tamil Nadu 600117, India*
[b]*Centre for Advanced Research and Development/Chemistry, Vels Institute of Science, Technology & Advanced Studies, Chennai, Tamil Nadu 600117, India*

subraghu_0612@yahoo.co.in, rakvani@yahoo.co.in

Energy conversion and storage have emerged as major global challenges as a result of rising energy demand. Layered materials with a crystal framework are analogous to graphite and have garnered extensive attention in energy-related applications. In particular, layered materials are prominent due to their exceptional performance in practical applications, which non-layer materials strive for. In addition, they exhibit distinctive properties such as proficient guest ion intercalation, rapid ion diffusion, unprecedented initial Coulombic efficiency, surprising structure reversibility, superior mechanical adaptability within

Layered Materials: Synthesis and Applications in Energy and Environmental Remediation
Edited by Alagarsamy Pandikumar, Kandasamy Jothivenkatachalam, and Subramanian Moscow

ISBN 978-981-5129-17-5 (Hardcover), 978-1-003-50831-1 (eBook)
www.jennystanford.com

the layers, and exceptional stiffness perpendicular to the layers. These properties are facilitated by the well-maintained in-plane integrity of the layers and attest them to be superior when compared to the non-layered materials. This book chapter gives amplified attention to the structural qualities of layered materials, synthesis, and electrochemical applications of less cobalt/cobalt-free layered lithium and manganese-rich oxides (LMRO) and investigates their unique performance in energy storage resulting from their inherent structural features. A quick overview of common layered materials and their crystal structures will be discussed followed by a summary of LMRO structure's enhanced characteristics, preparatory methods, and eco-friendly redemption and significance.

5.1 Introduction

The fast-moving world's most indispensable need is energy storage. The need for powerful energy storage systems with excellent power density and a better cycle life is exceptionally increasing. In order to encounter this challenging demand, serious efforts have been rigorously employed to enhance the electrode performance to achieve significantly improved capacity, high C-rate, and longer cycle life. Layered materials are a broad category with a wide range of forms and functions. The two main types of layered materials are natural (or crystalline) and artificial (or synthetic). Despite being quasi-crystalline materials in nature, the former layered structures are generated by particular designs, whilst the latter structures are formed by unique crystal properties. As unit building blocks, all layered materials have two-dimensional (2D) closely connected slabs (so-called host layers) made up of covalently coupled atoms (e.g., graphite, h-BN, g-C_3N_4) or edge/corner/face-shared $MX6$ octahedral (e.g., M = transition metal; X = O, S, Cl, C, etc.). Because of their strong in-plane chemical connections, weak intermolecular forces between neighboring layers, strong covalent bonding inside the intra-layer, high reactivity, wide operating voltage window, enhanced performance via the formation of extra chemical bonds, facile phase modulation, encouraged functional diversity, remarkable flexibility, effective strain management, and capacity to function as ideal supporting material in electrocatalysis, layered materials have been popular in recent years [1, 2].

5.1.1 Types of Layered Materials in General

Novel layered materials are classified into three groups based on the overall structure of the layers and the tensions that exist between the layers that pack the layers together.

Type 1: Thin sheets of atoms bound together by van der Waals force fall under this category. Graphite and boron nitride are the best examples of type 1 layers. Both electron donors and acceptors have been observed intercalating in graphite, and the intercalation mechanism is predominantly electronic. Carbon layers that are positively or negatively charged may form depending on the guest species. Graphite intercalation compounds are uncommon among layered host materials because they have a high degree of ordering.

Type 2: Dichalcogenides commonly denoted as transition metal dichalcogenides (TMDs) come under this category. The structural network is formed as a layer made up of a metal atom sheet sandwiched between two sheets of chalcogen atoms X = S, Se, or Te. Layered TMD becomes a very suitable irreplaceable active material for electrochemical energy storage for the following prominent aspects:

- ✓ There is a wide contact area between the active species and the electrolyte, due to the increased specific surface area, allowing for rapid charging/discharging reactions at the surface of stacked TMDs.
- ✓ The under-coordinated edge sites of multilayer TMDs can operate as metal ion adsorption sites, resulting in increased metal-ion storage capacity.
- ✓ The interlayer space between the layers easily permits fast ion diffusion, insertion/de-insertion, and accommodation during the metal-ion insertion process since the adjacent X-M-X layers are generally neutral and separated by a van der Waals gap.

Layered TMDs usually possess three polymorphs. They are trigonal phase, hexagonal phase, and rhombohedral phase. TMD polymorphs can be easily turned from one phase to another and the electrochemical performance varies accordingly. For example, MoS_2 is a hexagonal semiconducting phase, which can be transformed into a trigonal metallic conducting phase under specific conditions. Aside from their respective relevance, layered TMDs are frequently hybridized with carbon materials, metal oxides, and conducting polymers, resulting in unexpected characteristics. Electron transport between host species and recipient layers is common in TMD intercalation. Because of the effective transport of electrons to host compounds, the insertion of metal atoms, particularly alkali metal atoms, results in unique electronic characteristics [3–6].

Type 3: Layer double hydroxides and layer silicates (clays) are examples of this. The condensation of sheets of connected Si $(O, OH)_4$ tetrahedra with sheets of linked $M_{2-3}(OH)_4$ octahedra yields layers. Between the two tetrahedral sheets is packed the octahedra sheet M, which can be a divalent or trivalent cation. Another alternative is condensation, which results in two sheet minerals with the general layer formula M_{2-3} $Si_4O(OH)_2$. Layered materials that come under type 3 can be further classified, according to their structure, into layered double hydroxide (LDH), hydroxy double salt (HDS), and layered hydroxy salt (LHS).

LDH has a basic structure similar to brucite [Mg $(OH)_2$], with Mg cations in the middle of the edge-sharing hydroxyl octahedra, forming an infinite 2D layer. The hydroxy groups are arranged in a triangular lattice that runs orthogonal to the plane of the layers. The real geometry around the metal, as well as the packing of hydroxyl anions, deviated from the idealized spatial space. The octahedra are compressed along the stacking axis, resulting in a geometry that is centered on the metal rather than the simplified O. Dispersion forces and hydrogen bonding are considered to be the driving factors that hold the layers together.

On the basis of the structure of copper hydroxy nitrate and zinc hydroxy nitrate, HDS and LHS have been categorized into two structural categories. HDSs are rigid, and their high-layer rigidity and structural anisotropy exhibit a variety of physical properties and structural phenomena that are both fundamental and practical. Other common applications of type-3 materials include selective anions exchange, topotactic insertion of divalent cations, metal oxide precursors, nanosheets, and templates for metal nanoparticles as shown in Scheme 5.1 [3, 7].

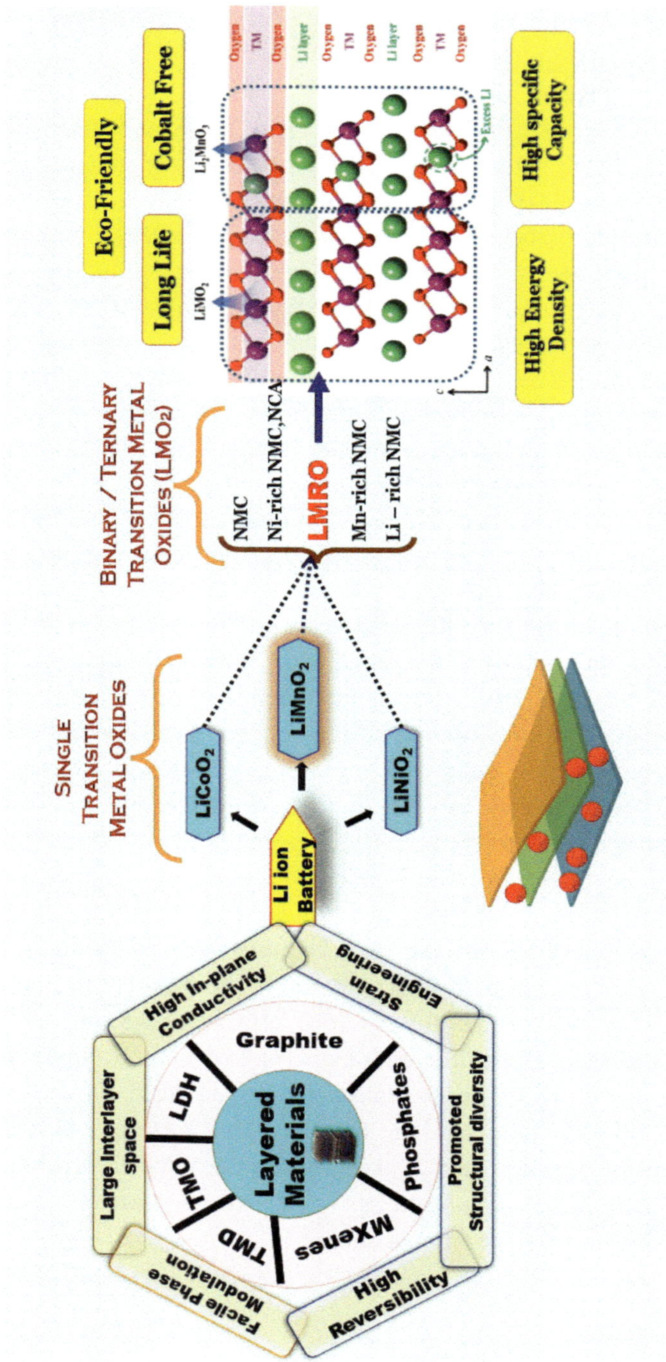

Scheme 5.1 Representing the summary of layered materials.

5.2 2D Layered Materials in Energy Storage Applications

Due to their versatile 2D designs and enormous theoretical capacities, LDH and layered metal oxides are regarded as suitable electrodes for the new development of advanced efficient secondary-ion batteries and supercapacitors. In 2010, a paradigm change occurred when the Nobel Prize was granted for the invention of graphene, a two-dimensional monolayer of carbon generated via exfoliation that defied graphite's physical and chemical properties. This attempt possibly made researchers have a great passion and quest in developing 2D layered materials, especially for applications related to energy storage. 2D layered materials can provide a sufficient number of active sites, effective diffusion channels, and unique features such as foreign ion insinuation, good mechanical strength, and improved electrochemical reactions. The most significant concept that generates intriguing qualities for enhanced electron conduction (or) phase transition and altering the chemistry interlayer environment is guest intercalation. Intercalation chemistry still stands as a greater challenge for the bulk layered material, leading to a depletion of active sites and inhibiting ion transportation. In general, intercalation can be defined as inserting/introducing the guest species into the interlayer space of the host framework. In the host materials, the layers are bonded in weak van der Waals interactions and voids are present in between the layers, where the guest species can be accommodated in the voids. By intercalation reaction, the properties of both host and guest species are expected to be modified/enhanced through guest–guest and guest–host interactions. From organic to inorganic, a wide spectrum of guest molecules can be used. One of the most common intercalation reactions is between common intercalation reaction that is between metal–metal intercalation, which has valid applications in metal-ion batteries and electrochemical energy storage devices [8].

5.3 An Overview of Layered Materials Employed in Lithium-Ion Battery

The idea for a lithium battery was first envisioned in the 1970s and began to see widespread adoption by the 1990s. When the limited energy densities and capacities of the developed batteries inspired the search for better configuration, lithium became a target. Lithium batteries can provide a high storage efficiency of 83% and become a possible power source for sustainable transport. It was assumed at the time that metallic lithium would serve as the anode in the batteries, so special attention was paid to identifying matching cathode materials. Following the studies on ionic conductivity in solids, materials with high reduction potential that could facilitate lithium ions at high transfer rates stimulated researchers' interest. As a result, a variety of lithium-containing structures were explored, and the material's behavior during alkali metal intercalation under reductive conditions was assessed. This was no easy task, as this material should ideally meet a number of prerequisites for subsequent, efficient incorporation in batteries [1]. Therefore, the matching cathode material should have:

- ✓ A widely available electronic band structure that permits a constant, large insertion-free energy change across a composite range.
- ✓ A capacity to accommodate the guest species over a wide stoichiometric range with negligible framework modification.
- ✓ A high alkali ion diffusion coefficient within the structure.
- ✓ A possibility for both faradaic and non-faradaic process.
- ✓ An efficiency to operate at near-ambient temperatures.
- ✓ An intractable nature of the electrolyte exhibits no co-intercalation of electrolyte components.

Here below some of the cathode materials of particular interest are discussed.

5.3.1 $LiTiS_2$

In 1965, titanium disulfide (TiS_2), a metal chalcogenide of the type MX_2 with potential lithium binding sites, was first found to host lithium ions. There were a few issues to sort out, but a workable rechargeable battery was demonstrated in 1976. The battery cell was built using Li metal as an anode, TiS_2 as a cathode, and $LiPF_6$ as an electrolyte. A cell electromotive force (emf) of 2.5 V was observed, with an initial current density of 10 mA/cm^2, indicating that the reaction was single-phase: xLi + $TiS_2Li_xTiS_2$. However, Li_xTiS_2 had a relatively low cell voltage of only 2 V. However, due to the creation of unexpected lithium dendrite, the highly reactive lithium metal could not be entirely controlled, posing a fire threat [9].

5.3.2 $LiCO_2$: LCO

The discovery of Li_xCoO_2 by John B. Good and his colleagues at Oxford University in 1979–1980 was a watershed moment (LCO). In comparison to Li^+/Li, CoO_2 has a high potential of 4–5 V. In 1991, a commercial lithium battery was released, as a result of continual discoveries and advances, having up to 4.1 V charging voltage with an energy density of 80 Wh/kg or 200 Wh/L. Gradually, graphite made its way into the field to serve as an anode. Batteries with 4.2 V charging voltages and energy densities of 400 Wh/L can be made with this anode material [10].

5.3.2.1 Major benefits

High power density, good electrical conductivity, high open circuit voltage, and minimal self-discharge are all advantages of Li_xCO_2. It was the first successful commercialization of a proven technique, which provided researchers with a better notion of how to proceed. The structure of LCO is well-ordered.

5.3.2.2 Major defects

LCO also has several drawbacks, such as a poor specific capacity of 140 mAh g^{-1}, which is only half of its calculated values, fluctuation caused by band overlaps between $Co^{3+/4+}$: t_{2g} and O^{2-}: 2p.

Co^{3+} oxidizes to Co^{4+} at high voltages, followed by the elimination of electrons from the O^{2-}: 2p band, causing phase changes in the crystal framework.

5.3.2.3 Alternative

Numerous strategies have been used, including carbon coating and coating with compounds such as Al_2O_3, $CePO_4$, $FePO_4$, $Li_4Ti_5O_{12}$, SnO_2, TiO_2, and ZnO_2, which have been tried to compensate for LCO's low capacity, and there appears to be a small increase in capacity and cycle life, but not to the expected level to meet the coming energy demand. Doping with metals including Al, Cr, Ni, Ti, and Mg, as well as the inclusion of Ca^{2+} ions, have all been tried (Table 5.1).

However, cobalt's high cost, relative scarcity, and toxicity encouraged researchers to seek out an ideal replacement transition metal element that minimizes overlap with the O^{2-}: 2p band while also providing ecological and economic benefits [10–12].

Table 5.1 Enumerates the significance of $LiCoO_2$ (LCO)

$LiCoO_2$ Innate nature		Specific capacity mAh g^{-1}	Potential V	Modifications any	Ref.
Structure: Layered					[13]
Theoretical capacity: 274					
Available capacity		140	3.8	-	
		185	3.95	-	
		190	4.45	-	
		215	4.55	-	
		200	4.6	Coated with $LiAlO_2$	[14]
		224.9	4.6	Doped with Mg	[15]
Operating voltage			3.9	-	[13]
Galvanometric energy density (Wh kg^{-1})	**740**		4.45	-	
	840		4.55	-	

5.3.3 $LiNiO_2$: LNO

In this search line, $LiNiO_2$ (LNO) layered material known to be isostructural with LCO, came into existence in 1954 by Dyer et al. Ni is fairly inexpensive and abundant.

5.3.3.1 Major benefits

It has a theoretical capacity of 275 mAh g^{-1} and an operating voltage of over 3.7 V when compared to Li^+/Li. When compared to LCO, it is less expensive, less toxic, and has a larger reversible capacity. Prior to oxygen depletion, the $Ni^{3+/4+}$: 3d band in $LiNiO_2$ just rests high above the top of the O^{2-}: 2p band. As a result, the preponderance of the lithium could be recovered from $LiNiO_2$, resulting in a higher capacity. The lattice oxygen does not alter even at a substantial electrostatic interaction level of Li^+ [11].

5.3.3.2 Major defects

- Because of the instability of Ni^{3+} above 2500 °C, the synthesis of high-quality $LiNiO_2$ samples with pure Ni^{3+} element at high temperatures is almost difficult.
- Concerns about its safety at high charge levels.
- Because their ionic radii are so comparable, cation exchange between Li^+ (0.76) and Ni^{2+} (0.69) lowers overall performance.
- The presence of two or more nickel atoms in the route of intercalation makes lithium-ion intercalation difficult.
- Due to the instability of nickel's +4 oxidation state, $LiNiO_2$ has low thermal stability and capacity fading.
- At various intercalation sites, LNO goes through a variety of phase transitions.

5.3.3.3 Alternative

Researchers were looking for ways that would lead to more reliable cathode materials, due to difficulties in synthesizing high-quality samples of $LiNiO_2$ and worries about its safety at a high state of charge (Table 5.2).

The majority of these solutions focus on replacing other elements for Ni in the layered structure, resulting in a sequence of

Ni-based oxides. Partially substituting Ni with Co, for example, has been proven to be an efficient way of reducing cationic disorder, and adding a minimal fraction of Al can increase both thermal and structural stability. The inclusion of Mn to the layered structure has been demonstrated to stabilize it. Several promising compositions, such as $LiNi_xMn_yCo_zO_2$ (NMC) and $LiNi_{0.8}Co_{0.15}Al_{0.05}O_2$ (NCA), have been identified using this element substitution technique and have become the most intensively explored cathode materials [12, 16–18].

Table 5.2 The electrochemical behavior of $LiNiO_2$ with respect to the synthesis route

$LiNiO_2$	Specific capacity mAh g^{-1}	Potential V	Synthesis route	Ref.
Theoretical capacity	274	2.2–4.6	–	[19]
Available capacity	190	4.2	Solid state	
	160		Acid assisted	[20]
	150		Microwave assisted	[21]
	170		Combustion assisted	[22]
Operating voltage	–	4.2	–	
Galvanometric energy density (Wh kg^{-1}): 629	–	–	–	

5.3.4 $LiMnO_2$: LMO

$LiMnO_2$ has been offered as an alternative electrode material to LCO because of its enormous theoretical capacity. Despite having the same chemical formula as LCO and LNO, $LiMnO_2$ crystallizes in an orthorhombic phase, as contrary to the layered O_3 structure (space group: Pmmm). $LiMnO_2$ is found in two different structures: monoclinic (layered, C2/m) and orthorhombic (zigzag layered structure, Pmmm), with the latter being more thermally stable. An ion exchange mechanism between Na^+ and Li^+ in layered $NaMnO_2$ has been shown to produce a layered structure of $LiMnO_2$.

5.3.4.1 Major benefits

$LiMnO_2$ has a higher power density and a higher capacity of 285 mAh g^{-1}. In addition, compared to LCO and LNO, $LiMnO_2$ is less expensive and safer for the environment.

5.3.4.2 Major defects

However, the minor difference in ionic radii between Li^+ (0.76) and high spin Mn^{3+} (0.65) makes it tough to stabilize layered $LiMnO_2$ and causes thermodynamic instability, resulting in poor insufficient electrochemical activity. $LiMnO_2$ has been shown to undertake a phase shift from a layered to a spinel structure ($LiMn_2O_4$), resulting in a reduced charge and discharge cycle life.

5.3.4.3 Alternative

$LiMnO_2$ material is converted into a more stable spinel $LiMn_2O_4$ framework during charging, resulting in rapid capacity reduction during cycling, due to the instability associated with the low spin Mn^{3+} ion (Jahn–Teller ion). Because both O_3-$LiMnO_2$ and $LiMn_2O_4$ share the same oxygen network, this phase transition is advantageous in Mn-derived $LiMnO_2$ phases, allowing for efficient cationic species transport.

The internal redox process starts when $LiMnO_2$ comes into contact with the electrolyte, and the Mn^{2+} produced simply diffuses from the Mn layer to the Li layer. The crystal structure of $LiMnO_2$ cathodes can be stabilized by replacing Mn with low valency metal ions, according to research [10, 11, 23, 24].

5.3.5 $LiMO_2$ (M = Ni, Co, Mn)

The focus and attention on single transition metal oxide cathode solutions have increasingly turned to mixed binary and ternary solid $LiMO_2$ solutions over time. Due to the complex chemistry of layered metal oxides, researchers investigated the effect of substituting various elements in $LiMO_2$ frameworks to achieve higher electrochemical performance and thermal stability. Compounds with α-$NaFeO_2$ structure and no transition metal ion migration in the lithium layers have outstanding electrochemical performance. The structural stability of layered oxides is

Table 5.3 LNMC cathode: A comparison of electrochemical performance with corresponding synthesis method and surface morphology

Cathode material	Synthesis route	Morphology	Initial discharge mAh g^{-1}	Capacity retention %	Ref.
$Li[Li_{0.144}Ni_{0.136}Co_{0.136}Mn_{0.544}]O_2$	Solid state	3D nanoporus	274	94.8	[27]
$Li_{1.2}Mn_{0.5}Ni_{0.13}Co_{.0.13}O_2$	Inverse microemulsion route assisted by a soft polymer	Micrometer-sized particles with flake-like morphology	200	93	[28]
$Li[Li_{0.2}Mn_{0.54}Ni_{0.13}Co_{.0.13}]O_2$	Co-precipitation	Micro grains	308	88.3	[29]
$Li_{1.2}Mn_{0.6}Ni_{0.14}Co_{0.06}O_2$	Self-ignition combustion	Interconnected pores-sponge	290	99	[30]
$Li_{1.2}Mn_{0.54}Ni_{0.13}Co_{.0.13}O_2$	Solid state	Macroporous	244	89	[31]
$0.5Li_2MnO_3$ $_0$.$5LiNi_{0.13}Co_{.0.13}Mn_{0.13}O_2$	Precipitation	Nanostructured bars	297.1	97	[10]
$Li_{1.2}Mn_{0.5\ 4}Ni_{0.16}Co_{.0.08}O_2$	Microwave hydrothermal		266.5	90	[32]
$Li_{1.2}Mn_{0.5\ 4}Ni_{0.13}Co_{.0.13}O_2$	Hydrothermal Method	Coralline-like hierarchical architecture	250.2	84	[33]
$Li_{1.2}Mn_{0.5\ 6}Ni_{0.12}Co_{.0.12}O_2$	Solvo hydrothermal	Tunable	261	85	[34]
$0.5Li_2MnO_3$-$_0$.$5LiNi_{0.13}Co_{.0.13}Mn_{0.13}O_2$	Co-precipitation	Hierarchically porous	262	83	[35]
$Li_{1.2}Mn_{0.5\ 6}Ni_{0.16}Co_{.0.08}O_2$	Microwave hydrothermal	Spheroidal shape with microporous	276.6	75.7	[36]

determined by the size of the transition metal ion in comparison to the alkali metal ion. This means that only Ni^{+3} and Co^{+3} are allowed to form 2D layered lithium-based oxides among the regularly utilized transition metal ions. Manganese substituted solid solutions, such as $LiNi_xMn_{1x}O_2$ (binary solid solution), have a steady high discharge capacity of 200 mAh g^{-1} because manganese doping improves thermal stability by reducing the massive exothermic reaction. Mn^{4+} ions also provide a very high energy barrier for cationic migration across tetrahedral sites, improving structural stability during cycling. NMC, Ni-rich NMC, Ni-rich NCA, Mn-rich, and Li-rich cathode materials are mixed transition metal oxide solid solutions that combine Mn, Co, and Ni and are widely referred to as NMC, Ni-rich NMC, Ni-rich NCA, Mn-rich, and Li-rich cathode materials. These cathodes can also be tailored to satisfy the requirements of applications [10, 11, 15, 25]. In addition to diverse dopants, the selective synthesis technique and distinctive morphology can adequately increase the electrochemical performance of the cathode materials [26]. The following is a list of NMC cathodes with various compositions (Table 5.3).

However, due to oxygen loss beyond 4 V, Ni-rich electrode materials like NCA and NMC suffer from structural and chemical instabilities, resulting in a significant self-heating rate below 200 °C, endangering lithium-ion cell safety. Over the last two decades, a concerted effort has been made to replace or substitute cobalt and nickel in layered oxides with manganese because lithium-ion cells with NMC cathodes offer competitive energy density and improved stability at high charging potentials (>4 V).

5.4 Cobalt-Free, Eco-Friendly LMRO in Trend

The high electrochemical capacity of over 280 mAh g^{-1} and high energy density of over 1000 Wh kg^{-1} of lithium and Mn-rich layered oxide (LMRO) cathode materials $xLi_2MnO_{3-(1x)}$ $LiMO_2$ (M = Co, Ni, Mn, Cr, etc.) have received a lot of attention because these are approximately double that of typical cathode materials. In the 1990s, the LMRO cathode materials found their way into the energy field.

In 1991, Thackeray et al. were the first to reveal that the lithium manganese oxide $Li_{2-x}MnO_{3-x/2}$, which is made from Li_2MnO_3, might be used as a LIB cathode material. The electrochemically active $Li_{2-x}MnO_{3-x/2}$ phases were obtained by chemical separation of Li_2O from the rock salt phase Li_2MnO_3 (Li_2O–MnO_2). The $Li_{2-x}MnO_{3-x/2}$ cathodes were found to have a discharge-specific capacity of roughly 200 mAh g^{-1}, according to the findings. In the same year, they observed that Li^+ ions can reversibly insert and withdraw from the MnO_2 host framework, resulting in a higher reversible electrochemical capacity of over 200 mAh g^{-1}. Thackeray et al. reported in 1992 that lithium manganese-oxide cathode materials generated from Li_2O–$yMnO_2$ (Li_2–MnO_3, $y = 1$) had a discharge-specific capacity of 213 mAh g^{-1} and had a remarkable electrochemical performance. For the first time, Kalyani et al. discovered in 1999 that the Li_2MnO_3 molecule may be electrochemically activated rather than acidically activated by charging in a lithium cell up to 4.5 V. At this high potential, the Li^+ ions can be removed from the Li_2MnO_3 phase, together with the release of oxygen from the electrode surface, resulting in a net loss of Li_2O from the structure. This finding represents a significant advancement in the development of LMRO cathode materials, meaning that the electrochemically inert Li_2MnO_3 material can be used as a cathode at higher operation potentials. Between the years 1999 and 2003, a number of Li_2MnO_3 combinations with various compositional ratios were investigated, and the electrochemical performance of the resulting cathodes was assessed for specific capacity more than 200 mAh g^{-1} at working voltages of 2–48 V vs Li/Li^+ as shown in Scheme 5.2. The structural integration of two components, as well as the integration of $LiMn_{0.5}Ni_{0.5}O_2$ with Li_2MnO_3, exhibited a truly solid solution, allowing researchers to better comprehend the LMRO structural framework. To explain the improved electrochemical performance of LMRO with discharge capacity 350 at higher cycle temperatures, Bruce et al. included the peroxide reaction and proton exchange processes into the LMRO reaction process and the same team also interpreted the oxygen release mechanism via in situ differential electrochemical mass spectrometry, a notion that stands for the rise of the specific capacity over the predicted value. Different approaches have been tried in the course of time to make the LMRO cathode material suitable for commercialization [10, 37, 38].

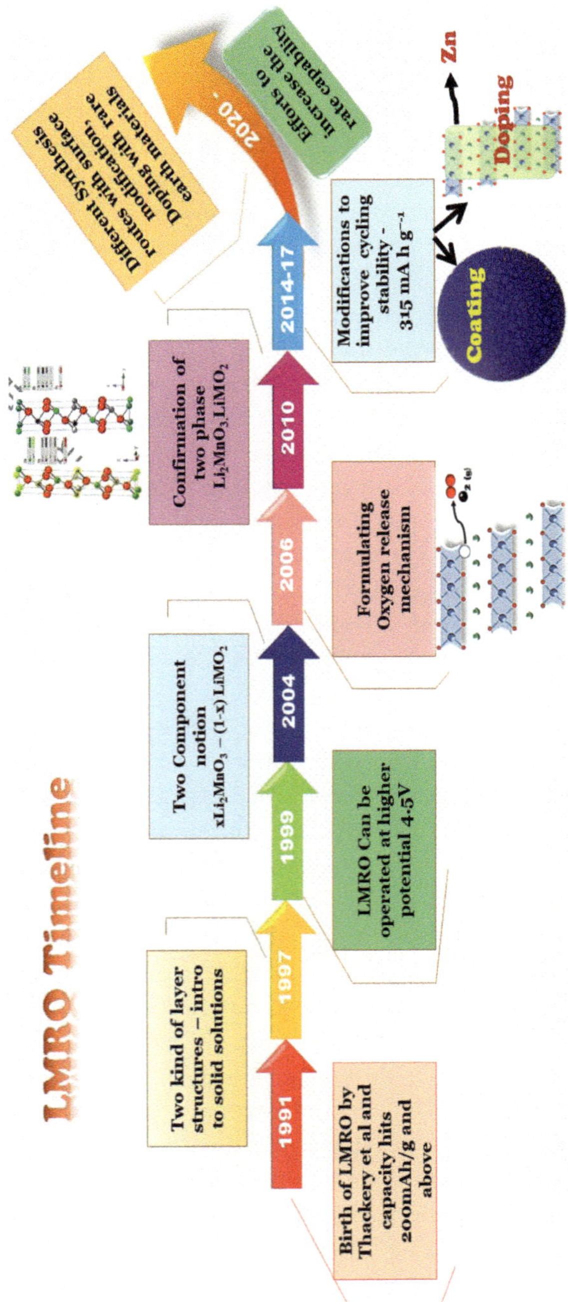

Scheme 5.2 Depicting the timeline of LMRO cathode materials.

5.5 Discussion on Structure of LMRO Cathode Material and Its Electrochemical Performance

The structure of LMRO still remains in debate. There are a lot of considerations in describing the structure of LMRO. Three critical points are being proved in every research article regarding LMRO's structure that is to be focused on now.

I. LMRO is generally thought to have a composite nature, consisting of two atomic structural phases. Monoclinic space group C2/m Li_2MnO_3 and trigonal space group R-3m $LiMO_2$. On a nanoscale-dominated nanocomposite structure, they appear to be intergrown. They are built like ordered layers of Li and transition metal alternately along the C-axis within the cubic close-packed oxygen array in this composite structure. When comparing LMRO to other typical layered oxides, the presence of Li at the Mn_6 superstructure unit is discovered to be a major structural difference.

II. The existence of a single solid solution phase with a monoclinic C2/m symmetry.

III. Finally, the single solid solution phase with trigonal (R-3m) structure.

In conclusion, there is no unified conclusion concerning the crystal structure of LMRO cathode material; the actual framework of the LMRO can be altered depending on the lithium elemental composition and synthetic conditions.

The structure of Li_2MnO_3 belongs to the R3m space group of the α-$NaFeO_2$ layered structure. The rock salt structure comprises several Li atoms in the 3a position, while the TM layer contains 1/3 lithium and 2/3 manganese in the 3b position, with the oxygen atom in the 6c position. Li_2MnO_3 with an O_3-type structure can also be mentioned as $Li[Li_{1/3}Mn_{2/3}]$, which is similar to the $LiMO_2$ structure. As a result, it is a distinctive example of $LiMO_2$ structure in which a single Li and two Mn atoms are positioned freely in the transition metal layer, resulting in a honeycomb-like pattern generated by the six Mn atoms surrounding each Li atom (Fig. 5.1).

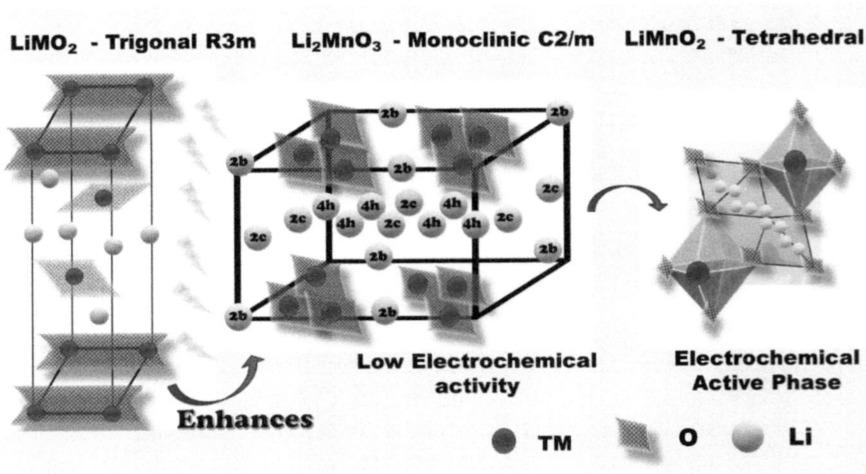

Figure 5.1 Structural representation of (a) $LiMO_2$ layered oxides, (b) Li_2MnO_3, and (c) $LiMnO_2$- honeycomb pattern.

The size, structure, and morphology of cathode materials are the most important factors affecting their electrochemical performance. Because LMRO is made up of two separate phases, trigonal $LiMO_2$ and monoclinic Li_2MnO_3, it can be difficult to distinguish between them. In this composite structure, Li_2MnO_3 is an electrochemically low active phase, yet it is a phase that increases the specific capacity of the LMRO cathode material. Li_2MnO_3 is converted to $LiMnO_2$, an electrochemically active phase activated by $LiMO_2$, after the first cycle. LMRO becomes a superior cathode material when compared to other cathodes due to the cooperative effect between Li_2MnO_3 and $LiMO_2$ [39].

The charging profile of LMRO can be understood in two parts below 4.5 V and above 4.5 V. De-intercalation of Li^+ from the $LiMO_2$ component occurs below 4.5 V, resulting in transition metal oxide redox. The behavior changes to Li_2MnO_3 above 4.5 V, assuming Li^+ de-intercalation from the Li_2MnO_3 component. This nature is only visible during the first cycle, after which it vanishes.

The electrode reaction in the first cycle, particularly oxygen loss, is blamed for the LMRO cathode's low initial Coulombic efficiency. The irreversible capacity loss observed during the first cycle could be attributed to the extraction of lithia (Li_2O)

from Li_2MnO_3, which results in the formation of a layered MnO_2 component ($Li_2MnO_3/Li_2O + MnO_2$) [37].

Lu and his colleagues presented an oxygen evolution mechanism to describe the first charge–discharge process in LMRO cathode materials. At 4.5 V, the simultaneous elimination of lithium and oxygen from the structure leads the cathode to lose capacity. The release of oxygen gas is assumed to be linked to structural changes during cycling, such as the creation of large pores, which causes TM redox couplings to shift to lower electron orbitals, resulting in a voltage decrease. The redox behavior of the TM and oxygen is discovered in the charging–discharging operation. Redox occurs in a voltage range of 4.8–3.5 V for Ni, Co, and O, and in the voltage range of 3.5–2.0 V for Mn [10].

The removal of Li^+ ions from the lattice below 4.5 V is accompanied by the extraction of electrons from the O_2-species near the surface. The Li^+ ion transfers from the transition layer to the lithium layer following the oxygen evolution, leaving voids that must be filled by transition metals migrating from the surface to the bulk. The process will continue until the lithium gaps are filled with transition metals and the oxygen evolution ends. Change metal migration into the Li layer is thought to cause the spinel phase transition of $LiMO_2$ (M = Ni, Mn, Co, R3m). The elimination of lithia (Li_2O), which generates lattice strain in the electrode, is linked to the phase change from Li_2MnO_3 (C2/m) to spinel. Transition metal redox couples are crucial for lithium-ion batteries to achieve substantial charge capacities [37, 40].

5.6 A Comparative Study on Synthesis/ Morphology/Electrochemical Performance of LMRO

The research study in this part is confined to focus on specific morphologies and their corresponding synthesis route that yields high specific capacity with a good rate capability. The surface and entire material morphology have a tremendous impact on increasing capacity and rate capability. The explanation for this is due to the porous nature of the metal oxides that enables Li ions to diffuse rapidly within the electrode contact, reducing the

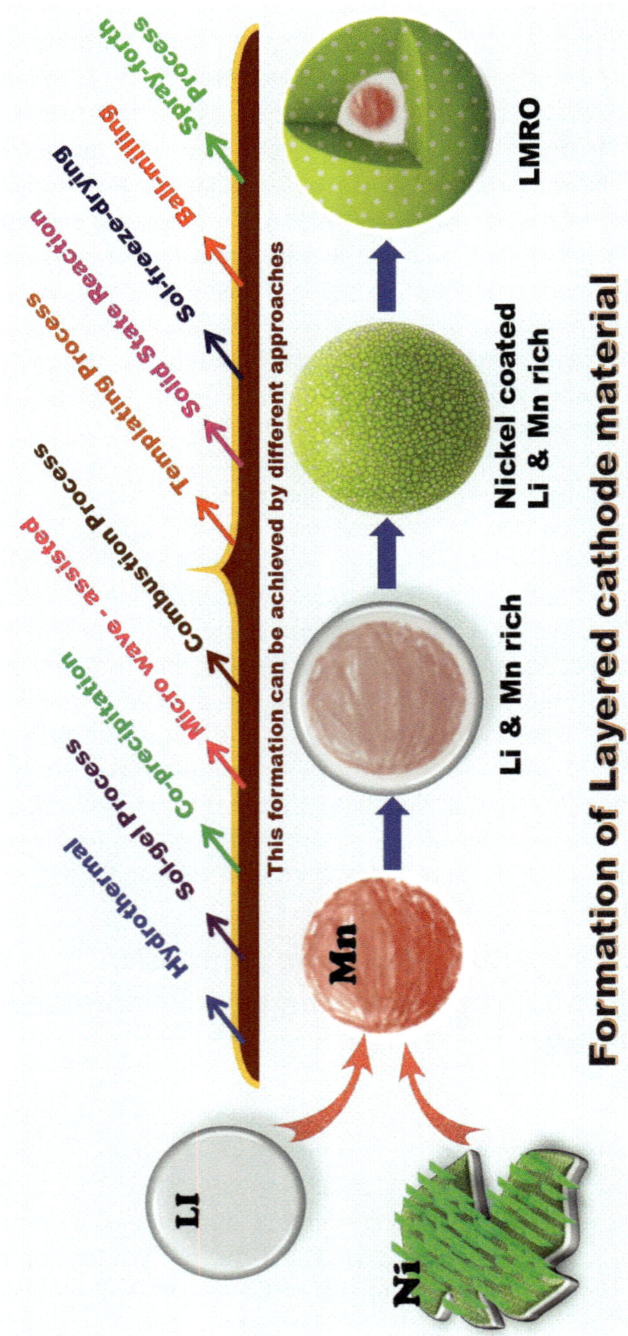

Scheme 5.3 Representation of the formation of layered cathode materials.

amount of slow solid-state migration. However, it is not only a question of whether the cathode material is porous or mesoporous; it is also a question of how the pores are interconnected and whether the associated linkages allow Li diffusion shortcuts. Li-ions should reach the interlayer channel in layered electrode materials, which are orthogonal to the basal plane. The synthesis procedure has a substantial influence on the crystal shape of LMRO and the resultant electrochemical performance. To standardize cobalt-free or low-cobalt LMRO of various ratios, a variety of approaches have been explored. The synthesis procedure has a substantial influence on the crystal shape of LMRO and the subsequent electrochemical performance. Cobalt-free LMRO of various compositions has been refined using a variety of approaches. The content in the following sections will focus on a few of the distinctive morphologies tailored to particular synthesis techniques, as well as the accompanying structural or surface alterations and electrochemical performance.

5.7 Hydrothermal Synthesis

One of the most prevalent synthesis processes for nanomaterial preparation is hydrothermal synthesis. This method can make nanomaterials that are unstable at high temperatures. Nanomaterials with a high vapor pressure can be created with minimal material loss using this process. By regulating the solution composition, reaction temperature, pressure, solvent characteristics, and maturing time, hydrothermal can effectively regulate grain size, crystalline phase, particle morphology, and surface chemistry [41]. In the preparation of LMRO cathode materials, a variety of hydrothermal techniques have been used, although just two are discussed here.

5.7.1 AlF_3-Coated Li-Rich Co-Free Mesoporous LMRO (Suppression of Voltage Fading)

Ashraf and his team have synthesized AlF_3, covering Li-rich Co-free layered $Li_{1.2}Ni_{0.2}Mn_{0.6}O_2$ as an effective means to reduce voltage fading using this technique. The samples were made using a hydrothermal method, which resulted in excellent crystallization in the layered structure. The AlF_3-coated $Li_{1.2}Ni_{0.2}Mn_{0.6}O_2$ has a capacity of 248 mAh g^{-1} and a voltage fading rate of 1.40 mV per cycle over 100 cycles. Hydrothermal synthesis was followed by solid-state calcination, with citric acid as a chelating agent, and AlF_3 coating was done via a chemical deposition technique. The AlF_3 coating, which was chosen as an inert material to coat the particles, reduces the loss of Li ions on the particles' surface, increases the Li_2MnO_3 component, and, as a result, increases the concentration of Mn^{3+} ions. When Mn^{3+} levels rise, Li_2MnO_3 gradually activates, resulting in a capacity boost during cycling. The morphology displays a hierarchical mesoporous structure with a size range of 4.5–5 nanometers. The structural and electrochemical stability of LMRO is greatly influenced by the conditions of synthesis and operating parameters [42, 43]. The mesoporous morphology of hydrothermally produced $Li_{1.2}Ni_{0.2}Mn_{0.6}O_2$ electrode materials had a role in increasing capacity. Another benefit of the AlF_3 coating is the lower operational voltage loss, which is half that of the pristine sample. Finally, this new venture has proved the efficiency of this coating in terms of structural stability, Mn^{3+} cation migration, suppression of cross growth in the spinel phase, and protection against electrolyte side reactions, contradicting certain assertions in the literature. It should be noted that the results were achieved by coating pristine $Li_{1.2}Ni_{0.2}Mn_{0.6}O_2$.

5.7.2 Rod-Like Hierarchical Nano/Micro Co-Free LMRO (Boosting Rate Capability)

To produce a rod-like hierarchical nano/micro $Li_{1.2}Ni_{0.2}Mn_{0.6}O_2$, a simple hydrothermal technique combined with solid-state calcination was applied. Two phases are involved in the manufacture of hierarchical structured lithium-rich materials. In step 1 of the basic hydrothermal process, the raw materials are combined into a rod-like micro-sized matrix made up of oxalate precursors with dense surfaces. Step 2 entails releasing gas to destroy the oxalate precursors, resulting in a large number of nanoparticles in the bulk matrix, which are then brought together by the fusion of the particles' edges. This material has a high discharge capacity of 266 mAh g^{-1} after 50 cycles at 0.1 C-rate. At a 1 C-rate, it has a maximum capacity of 224 mAh g^{-1}, while at a 2 C-rate, it has a maximum capacity of 214.3 mAh g^{-1}. The higher rate capability and cyclability are due to hierarchical nano/micro designs. Reducing lithium-ion diffusion routes and enhancing particle active surface areas by reducing particle size to nanoscale levels have been suggested as a possible strategy for improving the rate capability of layered oxides. This unique approach was used to create rod-like hierarchical nano/micro $Li_{1.2}Ni_{0.2}Mn_{0.6}O_2$ materials with sufficient discharge capacity, rate, and cycle. RH-LNMO materials' outstanding performance is due in large part to hierarchical nanostructures/microstructures as well as exceptional structural stability. Nanoparticles provide short diffusion lengths that facilitate rapid lithium-ion intercalation/deintercalation rates and electron transfer, while the micro matrix maintains a basic structure that is thermodynamically stable. As a result, hierarchical nano/micro $Li_{1.2}Ni_{0.2}Mn_{0.6}O_2$ materials could be attractive options for lithium-ion battery cathodes with good performance [44].

5.8 Co-Precipitation Method

Co-precipitation is a straightforward, cost-effective, and commercially viable process. It can be utilized to create a variety of shell materials on various core types. High yield, product purity, absence of need for organic solvents, easy reproducibility, and low cost are advantages of this method over other methods; nevertheless, the size, shape, and compositions are dependent on the reaction parameter.

5.8.1 Co-Free Hollow Microspherical LMRO with Regular Porosity (Rate Capability)

Shumei Dou et al. used a co-precipitation method followed by high-temperature calcination to construct hollow microspherical $Li[Li_{0.4}Ni_{0.38}Mn_{0.38}]O_2$. The morphology was carefully regulated in order to produce a cathode with improved electrochemical performance, notably rate capability. Micro-sized spherical particles have a number of advantages over irregular particles, including improved fluidity and dispersion, a high tap density that prevents side reactions, and the ability to prepare cathode materials. The hollow microspherical structure of $Li[Li_{0.4}Ni_{0.38}Mn_{0.38}]O_2$ can be seen in SEM pictures. Microspheres of 100–200 nm are seen under high magnification, forming porous surface structures. Microspheres have a hollow spherical structure with an inner diameter of ca. This technique has three main advantages:

I. The hollow microspherical porous structure, which is made up of large-surface-area nanoparticles, allows for a shorter Li-ion diffusion length, which results in enhanced electrochemical kinetics.
II. The internal vacuum space, along with the porous shell, provides plentiful and hierarchical pathways for easy electrolyte penetration into the active material.
III. Finally, hollow microspherical $Li[Li_{0.4}Ni_{0.38}Mn_{0.38}]O_2$ might dramatically increase structural integrity by partially alleviating the mechanical strain of recurrent Li-ion intercalation/deintercalation processes during cycling, thereby contributing significantly to good cycling stability.

To summarize the synthesis method's impacts, the sample has a typical a-$NaFeO_2$ layered structure with a weak reflection corresponding to $LiMnO_6$, a low degree of cation mixing, and hollow microspheres with porous surfaces. After 51 cycles at a current of 170 mA g^{-1} in the potential range of 2.5–4.8 V, hollow microspherical $Li[Li_{0.4}Ni_{0.38}Mn_{0.38}]O_2$ demonstrated a high discharge capacity of 213 mAh g^{-1}, with oxygen release and electrolyte side reactions happening only in the first charging operation [45].

5.8.2 Co-Free Microporous Spherical Ball LMRO

In a recent trend, a quick combustion-assisted co-precipitation preparation of Li $[Li_{0.1}Ni_{0.3}Mn_{0.7}]O_2$ reveals an exceptional electrochemical performance. Accelerated ignition using organic fuels at varied temperatures followed the co-precipitation approach. Even at a high voltage (4.8 V), the fine-tuned electrode maintained steady cycling with a capacity retention of 98.2% after 100 cycles. At 0.1 C, 0.2 C, 0.5 C, 1 C, 2 C, and 5 C, the maximal electrochemical rate capabilities are 325, 314, 259, 228, 173, and 109 mAh g^{-1}, respectively. It emphasizes structural alteration through the use of a novel metal oxide (1:3:7) ratio. The improved Li-ion intercalation/de-intercalation sites on the cathode material were mostly due to the fluctuation in processing temperature, which was substantially responsible for the noteworthy specific capacity/energy density and cycle stability of Mn-rich cathode materials utilizing simple processes. An increased surface area has a major impact on the electrochemical characteristics of the samples. The porosity of the sample has an impact on ion diffusion kinetics, which is a major factor in rate capacity performance, based on pore inter-connectivity. The material's structural structure and texture were revealed using FE-SEM pictures. The sample treated at 900 °C developed a spherical ball morphology with a microporous nature that improved electrochemical characteristics for battery applications [46].

5.9 3D Reticular LMRO Cathode Material by Hard-Templating Route

Using ordered mesoporous silica as a hard template, a hard-templating procedure was designed to generate a 3D reticular $Li_{1.2}Ni_{0.2}Mn_{0.6}O_2$ cathode material. $Li_{1.2}Ni_{0.2}Mn_{0.6}O_2$ is made up of two interlaced 3D networks and a mesoporous channel system in three dimensions. The material's huge specific area boosts rate performance while also improving solid-liquid contact in the final electrodes. 364.55 mAh g^{-1} and 259.14 mAh g^{-1} are the initial charge and discharge capabilities, respectively. Its function-promoting structure, which is based on a nano-3D network and a mesopore channel system, contributes to the material's outstanding performance. Nanoscale lithium channels, a large specific surface area, excellent thermal and mechanical stability, and easy access to the material's core characterize this structure. This study implies that 3D reticular $Li_{1.2}Ni_{0.2}Mn_{0.6}O_2$ as a cathode for future lithium-ion battery development is a viable option. In addition, the hard-templating technology used in this study could be used to increase the cycle stability and rate capability of lithium-ion cathode materials in general [47].

5.10 Sol-Gel Method

The sol-gel technique is widely used to change substrate surfaces. The most notable benefit of the sol-gel process is that it produces a large surface area and stable surfaces. The sol-gel process is a bottom-up approach that converts atoms into nanoparticles. Two key reactions are involved in the sol-gel method: (i) hydrolysis of the precursor in acidic or basic media, followed by (ii) copolymerization of the hydrolyzed products.

5.10.1 Hybrid of Spinel and Layered LMRO

The electrochemical performance of a $Li_4Mn_5O_{12}$–Li_2MnO_3 composite material was investigated in this study using the sol-gel process. The presence of hybrid spinel and layered structure is revealed by structural study using X-ray diffraction, transmission electron microscopy, and Raman measurement. The appearance of elongated spherical form particles in the $Li_4Mn_5O_{12}$–Li_2MnO_3 composite shows that the particle's average size was around 80 nm. The cubic $Li_4Mn_5O_{12}$ and monoclinic Li_2MnO_3 phases are readily seen in the XRD pattern. $Li_4Mn_5O_{12}$–Li_2MnO_3 material has a discharge capacity of 209.2 mAh g^{-1} and a capacity retention of 96.54% up to the 50th cycle at 0.1 C. The electrochemical results show that the $Li_4Mn_5O_{12}$–Li_2MnO_3 cathode has a high specific capacity, cyclic stability, and rate capabilities. The integration of spinel–layered hybrid structures has increased the electrochemical performance of the $Li_4Mn_5O_{12}$–Li_2MnO_3 composite cathode. In the integrated hybrid $Li_4Mn_5O_{12}$–Li_2MnO_3 composite cathode material, the spinel phase provides superior rate capability while the layered Li_2MnO_3 structure provides high capacity [48].

5.11 Eco-Friendly Redemption and Significance

The emergence and dominance of lithium-ion batteries in rapidly growing industries such as consumer electronics, electric vehicles, and renewable energy storage are driving enormous interests and investments in the battery sector. However, according to the statistics, even a rechargeable battery is end-of-life over time and ultimately all lithium-ion batteries will cease to work. Depending on chemistry, a variety of materials (e.g., Co, Ni, Mn, Li, Cu, and Al), configurations, and purposes, a lithium-ion can perform between 500 and 10000 cycles of charging and discharging. In this scenario, research activities have usually focused on the recovery of cobalt, nickel, manganese, and lithium elements, which are main inputs from $LiCoO_2$, NCA, $LiFePO_4$, and $LiMn_2O_4$ battery-grade materials by pyro-, hydro-, and bio-metallurgy processes.

- Pyro-metallurgy methods may not be favorable due to high energy consumption, the production of toxic gases (such as sulfur dioxide and carbon dioxide), and other factors.
- Bio-metallurgy methods employing Cu^{2+} as a catalyst were explored by Zeng et al.
- Hydro-metallurgy processes often consist of an acid-leaching step and a separation or recovery step.

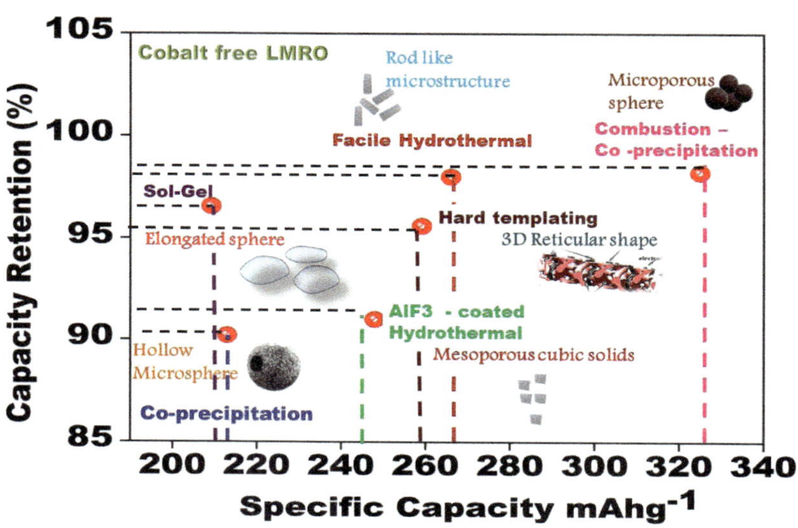

Figure 5.2 Summary of different synthesis techniques with corresponding electrochemical behavior.

Because of the great efficiency and low harmful gas emissions to the atmosphere, the major LIB recycling method researched and documented in businesses and academia is hydrometallurgical. Acid leaching is the most cost-effective, straightforward, and ecologically benign hydrometallurgical process for recycling used LIBs. Inorganic acids including HCl, HNO_3, H_2SO_4, and H_3PO_4, and organic acids like citric, DL-malic, ascorbic, succinic, oxalic, L-tartaric, and formic acids are commonly used as leaching agents for cathode materials from used LIBs. During the leaching of inorganic strong acids, Cl_2, SO_3, and NO_x are released, posing ecological damage and human health. Organic acids, in contrast to inorganic acids, are nearly natural and readily degradable, and some organic acids can be utilized as a reductant, precipitant, or complexing agent during the cathode material recovery process. In terms of leachate, the majority of study has concentrated on the separation of various metals, with a few papers looking into the reconstruction of new materials. However, hydrometallurgy and pyrometallurgy techniques have made real strides in recycling used LIBs, but there are still issues like slow workflows, low-value-added products, and major contamination. A fresh unique idea for recovering functional materials from used LIBs has been developed in recent years. There are two methods for material regeneration. One option is to repair used LIB cathode materials directly, while the other is to use hydrometallurgy leachate as a raw material in resynthesized materials.

Aside from LIB material regeneration, the creation of additional functional materials such as catalysts, sorbents, magnetic materials, and so on has steadily gained interest. The technique of recovering functional materials from expended LIBs is currently comparable to that of recovering cathode material. Depending on the synthesis of functional materials, several processes such as sol-gel, oxidation-reduction, and hydrothermal treatment can be used.

5.11.1 Function as a Catalyst

Due to their catalytic effectiveness in decomposing organic volatiles, manganese-based catalysts produced from used LIBs have sparked widespread attention. The advantages of regenerated manganese-based catalyst materials include always large specific surface area, plentiful mesoporous structure, and Mn^{4+}/Mn^{3+}

ratio. One of the most plentiful metals in the Earth's mantle is manganese; however, it is highly dispersed and widely distributed. The lithium-ion manganese oxide battery ($LiMn_2O_4$, Li_2MnO_3, or LMRO) is the type of Li-ion battery that does not contain cobalt. LMRO/NMC blend batteries are used in BMW and Nissan vehicles. LMRO batteries are environmentally friendly, economical, lack secondary pollution, low energy consumption, high kinetics, selectivity, and high efficiency. However, recycling of LMRO batteries is low-profitable due to the low price and high market availability of manganese compounds [49].

Reclaimed $LiMnO_4$ from used LIBs was utilized as a cathode material for sodium batteries, according to Nie et al. The Coulombic efficiency of the regenerated material was good. Li et al. used a hydrothermal technique to regenerate a spinel lithium-ion sieve ($Li_{1.6}Mn_{1.6-x}Fe_xO_4$ or LMO) from used LIBs. The LMO has been discovered to have a quicker ion diffusion advantage. As a result, it can be utilized to make the lithium-ion sieve and extract Li from the Salt Lake brine.

Cheng et al. generated a sea-urchin-like MnO_2–$NiCo2O_4$ anode material from discarded $LiNi_{0.6}Co_{0.2}Mn_{0.2}O_2$. Furthermore, the method's reasonableness was confirmed by the great results. A possible application direction is the regeneration of functional materials from exhausted LIBs [50, 51].

5.12 Conclusion

This chapter of the book provides an overview of typical layered materials and focuses on some of the pioneering material diagnostic research in the LMRO cathode. The LMRO cathode material has been proposed as a potential cathode material for next-generation high-energy-density LIBs, particularly in the realm of electric vehicles. However, various scientific concerns and challenges must be surmounted before commercial applications may be realized. This research also looked into certain synthesis methods for LMRO cathode materials that are completely free of cobalt. When compared to Mn, which is abundant and has low cost, Ni's low amount and the total absence of cobalt, both of which are harmful in nature, extremely expensive, and scarce.

This publication offers a summary of the materials regeneration process using wasted LIBs. It was discovered that wasted LIBs can be used to regenerate cathode materials and other functional materials, which has both environmental and economic benefits. Regenerated cathode materials from spent LIB leaching solutions are more practicable than directly repaired materials.

Despite the significant progress made, there are still some concerns that need to be addressed. It is crucial to build more environmentally friendly and cost-effective solutions for material regeneration [52].

References

1. Y. Guo, Y. Wei, H. Li, T. Zhai, Layer structured materials for advanced energy storage and conversion, *Small.* 13 (2017) 1–22. https://doi.org/10.1002/smll.201701649.
2. C. Wang, L. Zhang, Z. Zhang, R. Zhao, D. Zhao, R. Ma, L. Yin, Layered materials for supercapacitors and batteries: Applications and challenges, *Prog. Mater. Sci.* 118 (2021) 100763. https://doi.org/10.1016/j.pmatsci.2020.100763.
3. R. Mokaya, Ion exchange | Novel layered materials: Non-phosphates, *Encycl. Sep. Sci.* (2000) 1610–1617. https://doi.org/10.1016/b0-12-226770-2/03951-x.
4. Q. Yun, L. Li, Z. Hu, Q. Lu, B. Chen, H. Zhang, Layered transition metal dichalcogenide-based nanomaterials for electrochemical energy storage, *Adv. Mater.* 32 (2020) 1–29. https://doi.org/10.1002/adma.201903826.
5. P. Kumar, H. Abuhimd, W. Wahyudi, M. Li, J. Ming, L.-J. Li, Review—Two-dimensional layered materials for energy storage applications, *ECS J. Solid State Sci. Technol.* 5 (2016) Q3021–Q3025. https://doi.org/10.1149/2.0051611jss.
6. Y. Oaki, Exfoliation chemistry of soft layered materials toward tailored 2D materials, *Chem. Lett.* 50 (2021) 305–315. https://doi.org/10.1246/cl.200747.
7. C. Machingauta, Synthesis, characterization and application of two-dimensional layered metal hydroxides for environmental remediation purposes. (2013). ProQuest Dissertations and Theses; Thesis (PhD), Marquette University, 2013.; Publication Number: AAT 3603459; ISBN: 9781303581014; Source: Dissertation Abstracts International, Volume: 75-03(E), Section: B.; 205 p.

8. Y. Zhang, E.H. Ang, Y. Yang, M. Ye, W. Du, C.C. Li, Interlayer chemistry of layered electrode materials in energy storage devices, *Adv. Funct. Mater.* 31 (2021) 1–30. https://doi.org/10.1002/adfm.202007358.
9. T.H.E. Royal, S. Academy, O.F. Sciences, *Lithium-Ion Batteries*, 50005 (2019) 0–13.
10. M. Akhilash, P.S. Salini, B. John, T.D. Mercy, A journey through layered cathode materials for lithium ion cells – From lithium cobalt oxide to lithium-rich transition metal oxides, *J. Alloys Compd.* 869 (2021) 159239. https://doi.org/10.1016/j.jallcom.2021.159239.
11. T. Kim, W. Song, D.Y. Son, L.K. Ono, Y. Qi, Lithium-ion batteries: Outlook on present, future, and hybridized technologies, *J. Mater. Chem. A.* 7 (2019) 2942–2964. https://doi.org/10.1039/C8TA10513H.
12. J. Xu, F. Lin, M.M. Doeff, W. Tong, A review of Ni-based layered oxides for rechargeable Li-ion batteries, *J. Mater. Chem. A.* 5 (2017) 874–901. https://doi.org/10.1039/C6TA07991A.
13. Y. Lyu, X. Wu, K. Wang, Z. Feng, T. Cheng, Y. Liu, M. Wang, R. Chen, L. Xu, J. Zhou, Y. Lu, B. Guo, An overview on the advances of $LiCoO_2$ cathodes for lithium-ion batteries, *Adv. Energy Mater.* 11 (2021) 1–29. https://doi.org/10.1002/aenm.202000982.
14. J. Xie, J. Zhao, Y. Liu, H. Wang, C. Liu, T. Wu, P.-C. Hsu, D. Lin, Y. Jin, Y. Cui, Engineering the surface of $LiCoO_2$ electrodes using atomic layer deposition for stable high-voltage lithium ion batteries, *Nano Res.* 10 (2017) 3754–3764. https://doi.org/10.1007/s12274-017-1588-1.
15. F. Wu, J. Maier, Y. Yu, Guidelines and trends for next-generation rechargeable lithium and lithium-ion batteries, *Chem. Soc. Rev.* 49 (2020) 1569–1614. https://doi.org/10.1039/c7cs00863e.
16. S.T. Myung, F. Maglia, K.J. Park, C.S. Yoon, P. Lamp, S.J. Kim, Y.K. Sun, Nickel-rich layered cathode materials for automotive lithium-ion batteries: Achievements and perspectives, *ACS Energy Lett.* 2 (2017) 196–223. https://doi.org/10.1021/acsenergylett.6b00594.
17. T. Li, X.-Z. Yuan, L. Zhang, D. Song, K. Shi, C. Bock, Degradation mechanisms and mitigation strategies of nickel-rich NMC-based lithium-ion batteries, *Springer Singapore*, 2020. https://doi.org/10.1007/s41918-019-00053-3.
18. P. Oh, S.M. Oh, W. Li, S. Myeong, J. Cho, A. Manthiram, High-performance heterostructured cathodes for lithium-ion batteries with a Ni-rich layered oxide core and a Li-rich layered oxide shell, *Adv. Sci.* 3 (2016) 1–8. https://doi.org/10.1002/advs.201600184.

19. P. Kalyani, N. Kalaiselvi, Various aspects of $LiNiO_2$ chemistry: A review, *Sci. Technol. Adv. Mater.* 6 (2005) 689–703. https://doi.org/10.1016/j.stam.2005.06.001.

20. Y. Wang, H.Y.S. Huang, An overview of lithium-ion battery cathode materials, *Mater. Res. Soc. Symp. Proc.* 1363 (2011) 43–48. https://doi.org/10.1557/opl.2011.1363.

21. B. Park, Y.J. Kim, J. Cho, Cathodes based on Lico02 and Lini02, *Lithium Batter.* (2009) 410–444. https://doi.org/10.1007/978-0-387-92675-9_14.

22. W. Liu, P. Oh, X. Liu, M.J. Lee, W. Cho, S. Chae, Y. Kim, J. Cho, Nickel-rich layered lithium transition-metal oxide for high-energy lithium-ion batteries, *Angew. Chemie - Int. Ed.* 54 (2015) 4440–4457. https://doi.org/10.1002/anie.201409262.

23. M.M. Thackeray, M.F. Mansuetto, D.W. Dees, D.R. Vissers, The thermal stability of lithium-manganese-oxide spinel phases, *Mater. Res. Bull.* 31 (1996) 133–140. https://doi.org/10.1016/0025-5408(95)00190-5.

24. Z. Lu, J.R. Dahn, Understanding the anomalous capacity of Li/Li[Ni[sub x]Li[sub(1/3−2x/3)]Mn[sub (2/3−x/3)]]O[sub 2] cells using in situ x-ray diffraction and electrochemical studies, *J. Electrochem. Soc.* 149 (2002) A815. https://doi.org/10.1149/1.1480014.

25. D. Mohanty, K. Dahlberg, D. King, et al. Modification of Ni-rich FCG NMC and NCA cathodes by atomic layer deposition: Preventing surface phase transitions for high-voltage lithium-ion batteries. *Sci. Rep.* 6 (2016) 26532. https://doi.org/10.1038/srep26532.

26. L. Prettencia, E. Soundarajan, A.M. Shanmugharaj, K. Raman, Combustion assisted synthesis of Mn rich, Co free electrodes for high performance Li-ion batteries, *Electrochemical Behaviour Experimental Method Result & Discussion.* 862 (2020) 113962.

27. B. Qiu, C. Yin, Y. Xia, Z. Liu, Synthesis of three-dimensional nanoporous Li-rich layered cathode oxides for high volumetric and power energy density lithium-ion batteries, *ACS Appl. Mater. Interfaces.* 9 (2017) 3661–3666. https://doi.org/10.1021/acsami.6b14169.

28. T. Rao Penki, D. Shanmughasundaram, A. V. Jeyaseelan, A.K. Subramani, N. Munichandraiah, Polymer template assisted synthesis of porous $Li_{1.2}Mn_{0.53}Ni_{0.13}Co_{0.13}O_2$ as a high capacity and high rate capability positive electrode material, *J. Electrochem. Soc.* 161 (2014) A33–A39. https://doi.org/10.1149/2.008401jes.

29. X. Zhong, B. Papandrea, Y. Xu, Z. Lin, H. Zhang, Y. Liu, Y. Huang, X. Duan, Three-dimensional graphene membrane cathode for

high energy density rechargeable lithium-air batteries in ambient conditions, *Nano Res.* 10 (2017) 472–482. https://doi.org/10.1007/s12274-016-1306-4.

30. M.N. Ates, S. Mukerjee, K.M. Abraham, A high rate Li-rich layered MNC cathode material for lithium-ion batteries, *RSC Adv.* 5 (2015) 27375–27386. https://doi.org/10.1039/c4ra17235c.
31. S. Ma, X. Hou, Y. Huang, C. Li, S. Hu, K.H. Lam, Polymer microsphere-assisted synthesis of lithium-rich cathode with improved electrochemical performance, *Ceram. Int.* 42 (2016) 4899–4910. https://doi.org/10.1016/j.ceramint.2015.11.173.
32. S. Shi, T. Wang, M. Cao, J. Wang, M. Zhao, G. Yang, Rapid self-assembly spherical $Li_{1.2}Mn_{0.56}Ni_{0.16}Co_{0.08}O_2$ with improved performances by microwave hydrothermal method as cathode for lithium-ion batteries, *ACS Appl. Mater. Interfaces.* 8 (2016) 11476–11487. https://doi.org/10.1021/acsami.6b01683.
33. E. Wang, C. Shao, S. Qiu, H. Chu, Y. Zou, C. Xiang, F. Xu, L. Sun, Organic carbon gel assisted-synthesis of $Li_{1.2}Mn_{0.6}Ni_{0.2}O_2$ for a high-performance cathode material for Li-ion batteries, *RSC Adv.* 7 (2017) 1561–1566. https://doi.org/10.1039/c6ra26077b.
34. Q. Liu, H. Zhu, J. Liu, X. Liao, Z. Tang, C. Zhou, M. Yuan, J. Duan, L. Li, Z. Chen, High-performance lithium-rich layered oxide material: Effects of preparation methods on microstructure and electrochemical properties, *Materials (Basel).* 13 (2020). https://doi.org/10.3390/ma13020334.
35. W. Li, Review—An unpredictable hazard in lithium-ion batteries from transition metal ions: Dissolution from cathodes, deposition on anodes and elimination strategies, *J. Electrochem. Soc.* 167 (2020) 090514. https://doi.org/10.1149/1945-7111/ab847f.
36. S. Ma, X. Hou, Y. Li, Q. Ru, S. Hu, K. ho Lam, Performance and mechanism research of hierarchically structured Li-rich cathode materials for advanced lithium–ion batteries, *J. Mater. Sci. Mater. Electron.* 28 (2017) 2705–2715. https://doi.org/10.1007/s10854-016-5849-7.
37. H. Pan, S. Zhang, J. Chen, M. Gao, Y. Liu, T. Zhu, Y. Jiang, Li- and Mn-rich layered oxide cathode materials for lithium-ion batteries: A review from fundamentals to research progress and applications, *Mol. Syst. Des. Eng.* 3 (2018) 748–803. https://doi.org/10.1039/c8me00025e.
38. M.M. Thackeray, J.R. Croy, E. Lee, A. Gutierrez, M. He, J.S. Park, B.T. Yonemoto, B.R. Long, J.D. Blauwkamp, C.S. Johnson, Y. Shin,

W.I.F. David, The quest for manganese-rich electrodes for lithium batteries: Strategic design and electrochemical behavior, *Sustain. Energy Fuels.* 2 (2018) 1375–1397. https://doi.org/10.1039/c8se00157j.

39. Y. Li, Z. Li, C. Chen, K. Yang, B. Cao, S. Xu, N. Yang, W. Zhao, H. Chen, M. Zhang, F. Pan, Recent progress in Li and Mn rich layered oxide cathodes for Li-ion batteries, *J. Energy Chem.* 61 (2021) 368–385. https://doi.org/10.1016/j.jechem.2021.01.034.

40. W. Hua, S. Wang, M. Knapp, S.J. Leake, A. Senyshyn, C. Richter, M. Yavuz, J.R. Binder, C.P. Grey, H. Ehrenberg, S. Indris, B. Schwarz, Structural insights into the formation and voltage degradation of lithium- and manganese-rich layered oxides, *Nat. Commun.* 10 (2019) 1–11. https://doi.org/10.1038/s41467-019-13240-z.

41. Y.X. Gan, A.H. Jayatissa, Z. Yu, X. Chen, M. Li, Hydrothermal synthesis of nanomaterials, *J. Nanomater.* 2020 (2020) 8917013. https://doi.org/10.1155/2020/8917013.

42. A. Eftekhari, Lithium-ion batteries with high rate capabilities, *ACS Sustain. Chem. Eng.* 5 (2017) 2799–2816. https://doi.org/10.1021/acssuschemeng.7b00046.

43. A. Abdel-Ghany, A.M. Hashem, A. Mauger, C.M. Julien, Lithium-rich cobalt-free manganese-based layered cathode materials for li-ion batteries: Suppressing the voltage fading, *Energies.* 13 (2020) 1–22. https://doi.org/10.3390/en13133487.

44. L. Zhang, B. Wu, N. Li, D. Mu, C. Zhang, F. Wu, Rod-like hierarchical nano/micro $Li_{1.2}Ni_{0.2}Mn_{0.6}O_2$ as high performance cathode materials for lithium-ion batteries, *J. Power Sources.* 240 (2013) 644–652. https://doi.org/10.1016/j.jpowsour.2013.05.019.

45. S. Dou, P. Li, H. Li, Hollow microspherical $Li[Li_{0.24}Ni_{0.38}Mn_{0.38}]O_2$ as cathode material for lithium-ion batteries with excellent electrochemical performance, *Int. J. Electrochem. Sci.* 13 (2018) 9654–9668. https://doi.org/10.20964/2018.10.47.

46. L. Prettencia, E. Soundarrajan, A. M. Shanmugharaj, R. A. Kalaivani, S. Raghu, Combustion-assisted synthesis of Mn-rich cathode for high performance Li-ion batteries. https://doi.org/10.1016/j.est.2022.104054.

47. L. Li, L. Wang, X. Zhang, Q. Xue, L. Wei, F. Wu, R. Chen, 3D reticular $Li_{1.2}Ni_{0.2}Mn_{0.6}O_2$ cathode material for lithium-ion batteries, *ACS Appl. Mater. Interfaces.* 9 (2017) 1516–1523. https://doi.org/10.1021/acsami.6b13229.

48. R. Veena, J. Anbu Dinesh, S. Raman, P. Panigrahi, N. Ganapathi Subramaniam, Li and Mn-rich $Li_4Mn_5O_{12}$–Li_2MnO_3 composite cathode for next generation lithium-ion batteries, *J. Energy Storage.* 24 (2019) 100754. https://doi.org/10.1016/j.est.2019.04.028.

49. P.G. Schiavi, R. Zanoni, M. Branchi, C. Marcucci, C. Zamparelli, P. Altimari, M.A. Navarra, F. Pagnanelli, Upcycling real waste mixed lithium-ion batteries by simultaneous production of rGO and lithium-manganese-rich cathode material, *ACS Sustain. Chem. Eng.* 9 (2021) 13303–13311. https://doi.org/10.1021/acssuschemeng.1c04690.

50. R. Xu, W. Xu, J. Wang, F. Liu, W. Sun, Y. Yang, A review on regenerating materials from spent lithium-ion batteries, *Molecules.* 27 (2022) 1–16. https://doi.org/10.3390/molecules27072285.

51. A. Thesis, *Li-Ion Batteries Via a Direct Recycling Process.* (2020).

52. L. Duan, Y. Cui, Q. Li, J. Wang, C. Man, X. Wang, Recycling and direct-regeneration of cathode materials from spent ternary lithium-ion batteries by hydrometallurgy: Status quo and recent developments economic recovery methods for lithium nickel cobalt manganese oxide cathode materials, *Johnson Matthey Technol. Rev.* 65 (2021) 431–452. https://doi.org/10.1595/205651320x15899814766688.

Chapter 6

Synthesis and Electro-Catalytic Application of *h*-Boron Nitride and Graphene Oxide and Their Composites

Ragu Sasikumar[a] **and Byungki Kim**[a,b]

[a]*School of Mechatronics Engineering,*
Korea University of Technology and Education,
Cheonan, Chungnam 31253, Republic of Korea
[b]*Future Convergence Engineering,*
Korea University of Technology and Education,
Cheonan, Chungnam 31253, Republic of Korea

byungki.kim@koreatech.ac.kr

Layered materials (LMs) are the talk of the town in recent times, and are present in the form of graphite, graphene oxide (GO), gallium sulfide (GaS), hexagonal boron nitride (*h*-BN), germanium selenide (GeSe), nickel hydroxide ($Ni(OH)_2$), molybdenum disulfide (MoS_2), and bismuth selenide (Bi_2Se_3). Among them, h-BN, GO, and their composites have significant importance in the nanotechnology field. They can be differentiated from bulk materials by their range of size, morphology, and unique physico-chemical properties. The synthesized LMs show unique properties, such as excellent photo-catalyst support, good electron transfer, superior electron mobility, high surface-active sites/

Layered Materials: Synthesis and Applications in Energy and Environmental Remediation
Edited by Alagarsamy Pandikumar, Kandasamy Jothivenkatachalam, and Subramanian Moscow

ISBN 978-981-5129-17-5 (Hardcover), 978-1-003-50831-1 (eBook)
www.jennystanford.com

areas, mechanical strength, and thermal/electrical conductivity. These unique properties make them suitable for a range of industrial applications, including photo-catalysts, electrochemical sensors, biosensors, and energy storage applications like piezoelectric nanogenerators, batteries, supercapacitors, and corrosion treatments. In the current review, the focus is more oriented on different synthesis methods of *h*-BN and GO and their applications in the environmental remediation of GO and *h*-BN materials and their composites. The prospective scope of use of these LMs in the field of the electrochemical sensor is discussed briefly. This review chapter will help the researchers to grow in-depth knowledge of the boundaries and scopes of different synthesis methods, and their wide ranges of applications.

6.1 Introduction

In recent years, layered materials (LMs), like graphene oxide (GO), MXenes, hexagonal-boron nitride (*h*-BN), graphitic carbon nitride (g-C_3N_4), and molybdenum disulfide (MoS_2) have significant demand in multidisciplinary research areas [1–10]. The tunable LMs exhibit significant potential in many applications, owing to their important properties like high carrier mobility [11], quantum Hall effect [12], superconductivity [13], excellent flexibility [14], superior thermal conductivity [15], and high transparency [16]. Among LMs, GO and *h*-BN are used in a wide range of applications with sensors [17], optoelectronics [18], catalysis [19], biomedicine [20], electronics [21], and energy storage conversion [22].

Graphene is called the "material of the future" because of its unique properties and is one of the most researched materials in the world [23]. Graphene contains only carbon atoms, and each carbon atom is attached with sp^2 hybridized orbitals, while GO is an LM kind of carbon allotropes having many functional groups including keto (C=O), hydroxyl (–OH), and carboxylic (–COOH) groups [24, 25]. GO is classified into many types based on the number of layers such as GO (one layer), two-layered GO (two layers), few-layered GO (3–5 layers), multilayered GO (5–10 layers), and graphite oxide (more than 11 layers) [26]. Because of the GO synthesizing method, it has a number of oxygen-attached functional groups, and it is responsible for its unique properties including hydrophilic in nature and easy preparation in water or organic solvent medium. The highly oxidized form of GO is used as an electric insulator with a 2.2 eV bandgap. At RT, graphene exposes an excellent thermal conductivity of 5×10^3 W m-K^{-1} and a good electron mobility >200,000 cm^2 V s^{-1}. Graphene has remarkable properties and is used in a wider range of multi-disciplinary applications, including interconnects, transistors, ultra-capacitors, stretchable electronics, biosensors, photovoltaics, thermo-electrics, optical applications, and electro-mechanical-systems. The graphene 2D system has more interest from researchers in the fields of materials science, chemistry, physics, engineering, and medicine.

2D hexagonal boron nitride (*h*-BN) properties have been developed by van der Waals heterostructures. In 2008, it was first isolated from its few-atomic-layer by mechanical exfoliation but still has properties like bulk *h*-BN, too-bright luminescence under UV, a broader bandgap, good resistance during the oxidation process, lower density, and good thermal conductivity. The exfoliated *h*-BN gets low dimensional quantum confinement from its nano level and delivers particular optical and electrical properties. It also improves its chemical and mechanical characteristics. Graphene and other 2D materials' electrical properties extremely improve when *h*-BN is used as a substrate. *h*-BN monolayer and functionalized *h*-BN also show some exclusive attention such as energy conversion and storage when it is used as an active layer. Table 6.1 explains important layered materials with their properties.

Table 6.1 Important layered materials and their properties

Material	Property
Graphene, *h*-BN	Insulator
Group 4B: Hf, Ti, Zr	Diamagnetic nature; Semiconducting and insulating properties
Group 5B: Nb, V, Ta	Semi-metallic or metallic, charge density wave, superconducting, antiferromagnetic, paramagnetic, and diamagnetic properties
Group 6B: W, Mo	Semiconducting
Group 7B: Re, Tc	Diamagnetic nature; Semiconducting property
Group 10B: Pt, Pd	Diamagnetic nature; Semiconducting property

In this current review, we discussed briefly the synthesis techniques and electrochemical applications of *h*-BN, GO, and their composites. Figure 6.1 displays the schematic illustration of various types of LMs.

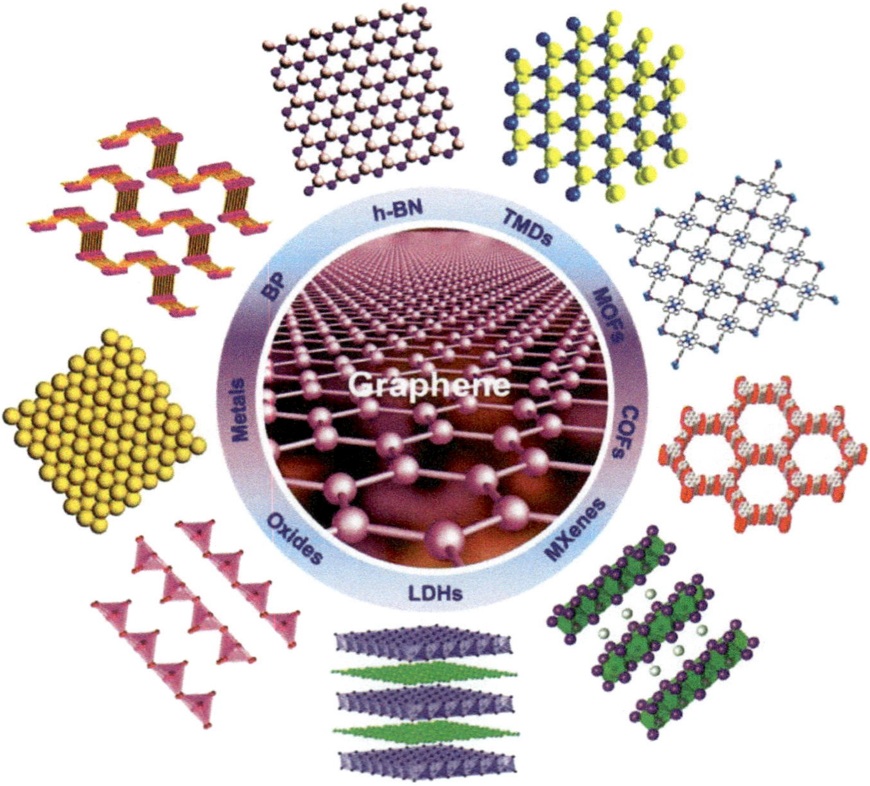

Figure 6.1 Graphic illustration of various types of LMs. (Reprinted with permission from Ref. [4]. Copyright 2015 American Chemical Society.)

6.2 Synthesis

6.2.1 Graphene Oxide (GO)

GO synthesis is mainly categorized into two main categories, namely, "top-down" and "bottom-up" methods. In the "bottom-up" method, the pristine graphene is built by the simple carbon molecules, and in the "top-down" method, the carbon source is used for extraction layers of graphene derivatives, usually graphite [27, 28]. The bottom-up synthesis is time-consuming and faces challenging methods like epitaxial development on silicon carbide wafers and chemical vapor deposition, etc. [29]. Therefore, the attention to the top-down method, the first produces GO, which is more popular use in nanocomposite materials. In 1859, B. C. Brodie (British Chemist) synthesized GO for the first time [30, 31]. This method is called the chlorate route, because, in this method, potassium chlorate ($KClO_3$) is used as an oxidizing agent. For the first time, B. C. Brodie treated graphite with many strong oxidizing agents to decode its structure. In B. C. Brodie's method, the graphite is treated with a mixture of HNO_3 and $KClO_3$ at 60 °C for 4 days and it is called Brodie's GO (BR-GO) [32, 33]. Due to the safer reactions, Hummers and Offeman made many improvements including $KMnO_4$ as an oxidizing agent, because $KClO_3$ releases toxic ClO_2 gas, and the addition of $NaNO_3$ in situ reaction to form HNO_3 relatively than using HNO_3 (solvent). Because of the safer and more accessible nature, the Hummers' method is generally used to generate GO. "Modified Hummers method" is a method proposed by Hummers by making improvements to GO synthesis, still this method is far from the standard method [34]. This is a method mostly used by researchers. In common, graphite flakes (carbon source) are mixed with phosphoric acid, sulfuric acid (a proton-donated solvent), and $KMnO_4$ (an oxidizing agent). Then, treated with H_2O_2 to eliminate the metal ions from the oxidizing agents, a yellow-brown liquid is obtained. After the dilute process, the semi-solids are treated with dil. HNO_3 to remove any metal ions. Then, washed with water by the centrifuge process until the pH of the solution is 7.0.

There are numerous methods to synthesize GO. The most general way is to use an oxidizing agent in an acidic environment. Other methods are electrochemical and microbial. There are some modified Hummers methods, such as two-step, nitrate-free [35], low- and room-temperature [36], and co-oxidant [37] methods. Then, in 2010, a novel method was proposed by Tour (own way) [38]. In this method, H_3PO_4 and H_2SO_4 (1:9 ratio) are mixed with $KMnO_4$, and graphite (6:1 equivalent weight) is added and placed in an ice bath and then heated at 80 °C and stirred for 12 h.

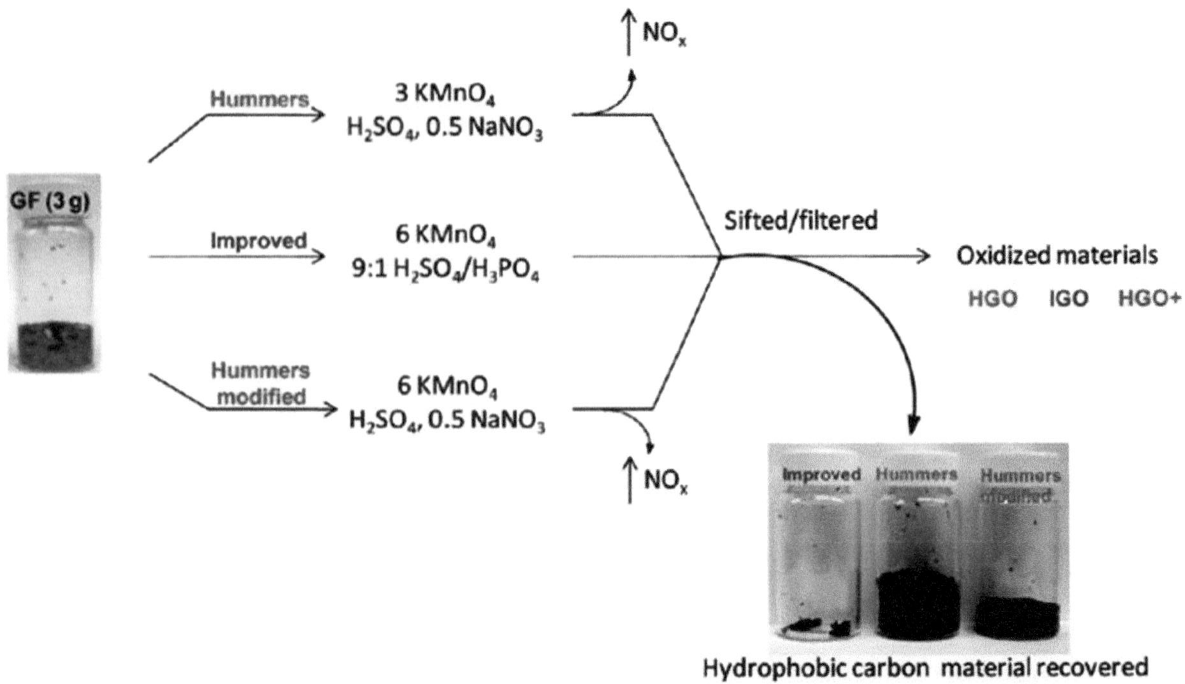

Figure 6.2 Diagram of most common synthesis methods of GO. (Reprinted with permission from Ref. [38]. Copyright 2007 American Chemical Society.)

Then, it is allowed to cool down to RT and poured onto ice. Finally, H_2O_2 (30%) is added in order to remove the excess $KMnO_4$. H_3PO_4 acts as a stabilizer of the oxidation process, which makes the synthesis of GO safe. This way yields a high quantity of GO with a more oxygen-containing functional group and normal regular structure. Figure 6.2 displays the most common synthesis methods of GO.

In 2018, Zhang et al. proposed a novel method of GO synthesis and they synthesized larger GO sheets with an approximate size of 108 μm by applying a standard oxidation method [39]. In Hummer's method, the synthesized GO is a significantly smaller size than its original nature in graphite carbon source. During the GO oxidation, the elastic nature produced by the addition of oxygen groups to the GO flake surface is due to the constant stirring during the mixing process. Without any stirring, they added H_2SO_4 and $KMnO_4$ followed by H_2O_2 to the above solution mixture, to produce comparable morphology and size of GO flakes to the source graphite. 3D structure of GO exfoliated under mild shaking or stirring. The larger GO sheets have significant properties and these are used for many composite materials.

6.2.2 Hexagonal Boron Nitride (*h*-BN)

From zero to three dimensions of bulk BNs, the different nanostructures of *h*-BN including QDs, NTs, and 2D NSs are synthesized [40, 41]. In this part, we discussed a few synthesis methods of *h*-BN. Chemical vapor deposition (CVD) is one of the common and effective synthesis methods for layered *h*-BN [42, 43]. At high temperatures, on the surface of substrates like Cu, Ni, Au, Pt (Table 6.2) [40–50], and Si/SiO_2, the gas-phase precursors undergo chemical reactions (or decomposition processes) to form 2D BN [44]. In this method, the general precursors like borazine (H_3BNH_3) [45], B-trichloroborazine ($B_3N_3H_3Cl_3$) [46], decaborane ($B_{10}H_{14}$)–NH_3 [47], N_2–H_2–BF_3 [48], and boric acid (H_3BO_3)–NH_3 [49] are used. Gaseous borazine and solid poly(aminoborane) are obtained when H_3BNH_3 is decomposed [50]. Hydrogenated α-BN is obtained when the temperature ranges from 125 and 200 °C. Layered *h*-BN is obtained when the temperature increases from 600 to 1100 °C.

Table 6.2 Preparation of layers of *h*-BN NSs on Cu, Ni, and Pt substrates

Precursor	Substrate	Conclusion
Borazane	Cu substrate; Chemical polishing or annealing at 1020 °C	High-quality *h*-BN layers with ~120 μm size
Borazane	Cu substrate; Annealing at 1000 °C	Mono and a few layers of *h*-BN obtained
Borazane	Ni substrate; Annealing at 800 °C	*h*-BN exhibited different morphologies depending upon various Ni grains
Borazane	Ni film; Annealing at 900 °C	*h*-BN with a thickness of ~50 nm obtained
Borazane	Pt substrate; Annealing at 1100 °C	A high crystalline layered *h*-BN was grown
Borazane	Pt foil; Annealing at 1100 °C	Uniform *h*-BN obtained

The quality and number of layers are controlled by altering the parameters, including pressure, temperature, precursor species, gas flux, substrate, and catalyst. For example, 3–6 layers of *h*-BN thin films on a Cu substrate were reported by Song et al. [51] and these films have great transparency in a broad wavelength range. By changing the ammonia borane concentration, the growth layers of *h*-BN on polycrystalline Pt can be obtained under normal pressures and reported by Cheng and co-workers found [52]. The most important issue for bulk synthesis of *h*-BN. Mechanical exfoliation happens when bulk crystals are broken down into smaller crystals by using different forces, including compression, scission during ball milling, and shear [53]. In these treatments, the interlayer space is improved by mechano-chemical methods (e.g., NH_3BH_3). Because of the poor yield of BN obtained by dry ball milling, strong sonication is necessary. To get more yield in wet mechanical milling, some solvents like NMP, sodium dodecyl sulfate, sodium hypochlorite, and benzyl benzoate [54] have been acting as exfoliating. In this case, to avoid layer structural problems in water and ethanol media, benzyl benzoate is an excellent viscous lubricant. Different structures of BN such as NSs, NTs, NRs, thin films, and NSs are displayed in Fig. 6.3.

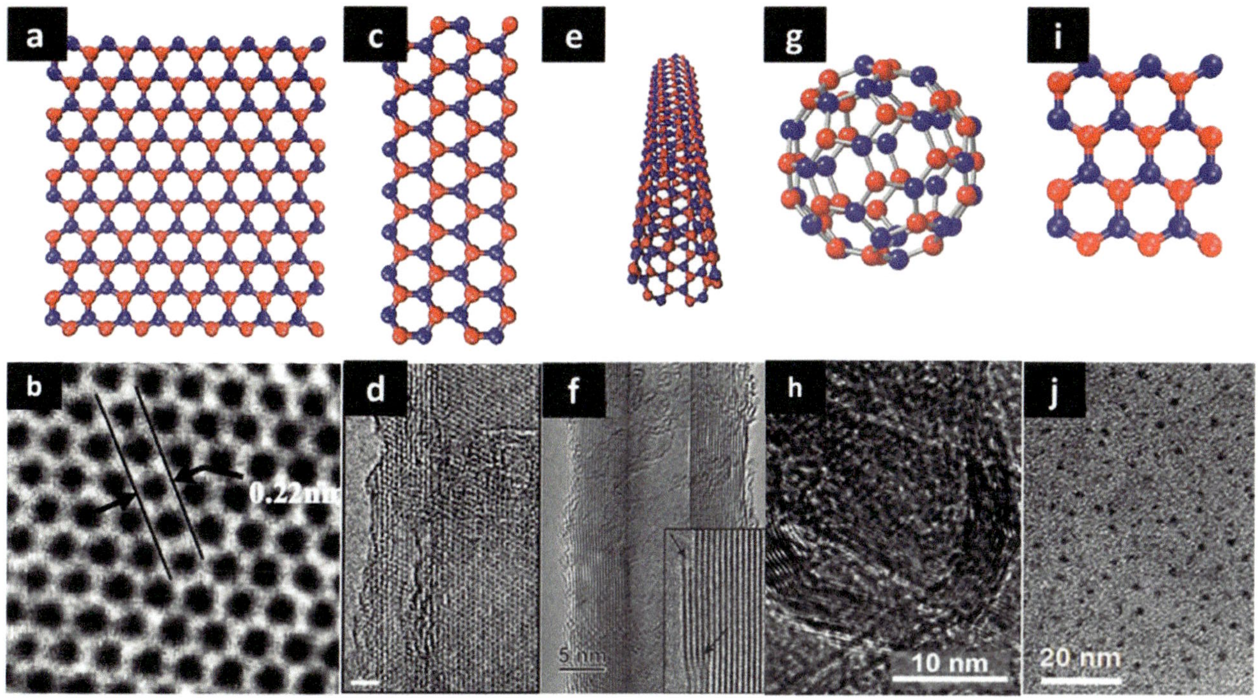

Figure 6.3 Different structures of *h*-BN: (a, b) NSs, (c, d) NRs, (e, f) nanorods, (g, h) fullerenes, and (i, j) quantum dots. (Reprinted with permission from Ref. [54]. Copyright 2011 Elsevier.)

6.3 Applications

GO, BN, and their composites have smart properties, such as tunable structural and morphological properties, excellent electron transportation, high surface area, high layer nature, high penetration, and high elasticity in functionalization or modification or functionalization, making them attractive for various types of applications. However, in this section, we discussed the potential applications of electrochemical sensors and nanogenerators.

2D-*h*-BN was reported as an active electro-catalyst with an Au electrode for oxygen reduction reaction by Uosaki et al. [55] and it is only potential on the Au electrode with no GCEs. In the after the above-reported work, Khan et al. described the 2D-*h*-BN as an electro-catalyst for the electro-catalytic detection of DA in the presence of UA quantification via the drop-casting method on the surface of SPEs [56]. The electro-catalytic performance was dependent upon the interaction between deposited material on the surface of GCE and electrolyte and comparison with unmodified GCE. Under optimized conditions, the peak determination between uric acid and dopamine was compared with other previously reported articles. Li et al. [57] reported a highly active 2D-h-BN/GCE composite for the electrochemical detection of DA and Vitamin C in the presence of UA, and it exhibited a good LR, LLOD, and better anti-interference capabilities.

There are many reports available on using 2D-*h*-BN with other nanomaterials/nanocomposite/hybrid nanocomposites. For example, Yola and Atar [58] reported electrochemical sensing of serotonin using 2D-*h*-BN doped with GQDs-modified GCE with molecularly imprinted polymers. This performance was exposed to electro-analytically enhanced outputs with already reported articles, due to high charge transfer capacities by the 2D-*h*-BN/GQDs, and the synergic effect between the GQDs and the 2D-*h*-BN [59]. Additional bare GCE is compared with 2D-*h*-BN/GQDs-modified GCE toward a redox probe, which shows considerable voltammetric peak response, and is changing to thin-layer adsorption type (at the 2D-*h*-BN and 2D-*h*-BN/GQDs-modified GCE) from diffusional (at the bare GCE). Later, it gives a higher peak response in terms of current, which is predictable [60]. The authors successfully used 2D-*h*-BN/GQDs-modified GCE for the real-time electrochemical performance of serotonin in human urine samples [61]. g-C_3N_4 doped 2D-*h*-BN-based nanocomposites are used to detect the methyl parathion [MP] [62]. These authors further did real-time sensing in orange juice samples for MP electrochemical detection. Similarly, 2D-*h*-BN/MWCNTs-modified GCE is used for the simultaneous electro-catalytic detection of β-agonists, ascorbic acid, and uric acid in human urine samples [63]. Other methods like graphene incorporated with 2D-*h*-BN nanocomposite for the electro-catalytic sensing of nicotine in tobacco samples [64] and 2D-*h*-BN/Bi_2O_3 nanocomposite for the electrochemical sensing of flutamide in environmental samples [65]. The synergic effect of mixing the many components including the nanocomposite is recognized as a useful electro-catalytic performance.

Shen et al. used *h*-BN as an electro-catalyst for the electrochemical sensing of 4-AP and phenol [66] and Pb [67]. The single precursor calcination method was used to prepare *h*-BN synthesis, as shown in Fig. 6.4, and the defect *h*-BN displays the occurrence of pore defects in the basal plane of D-*h*-BN, and these pore holes are active and produce more active sites in terms of chemically and electrochemically via bandgap sub-levels. In the above cases, the bare GCE electrochemical response was compared with D-*h*-BN to give excellent electro-catalytic performance, due to the *h*-BN showing higher surface area and good electron transportation, which was induced by the pore holes of the D-*h*-BN. The authors performed the effective electrochemical real-time detection of 4-AP and phenol in lake and tap water [66].

By using polymeric precursor methodology, Luo et al. [67] prepared *h*-BN whiskers. In this method, the authors used raw materials including melamine and boric acid in a tube furnace and slowly increased the temperatures up to ~1273 K in a flow of N_2/H_2 (5% H_2) atmosphere. Figure 6.5 shows the *h*-BN whiskers, which are ~500 mm in length and ~3 μm in diameter. The authors used this material as an electro-catalyst for nitrite. Figure 6.5(g) displays CV curves of *h*-BN/GCE compared to bare Ti electrodes.

Due to higher dispersibility (both water and solvents) with a good active site of GO, GO-based materials have a high range of active species (functional groups) and it makes excellent electro-catalytic and industrial applications. GO and GO-based composites displayed good thermal and electrical conductivity, good chemical stability, and outstanding electro and photocatalytic properties. GO-based composites have been used to develop various modified electrodes for many electro-catalytic applications, such as electro-catalytic sensors, electro-catalysis, electrochemiluminescence, biochemical sensors, and others.

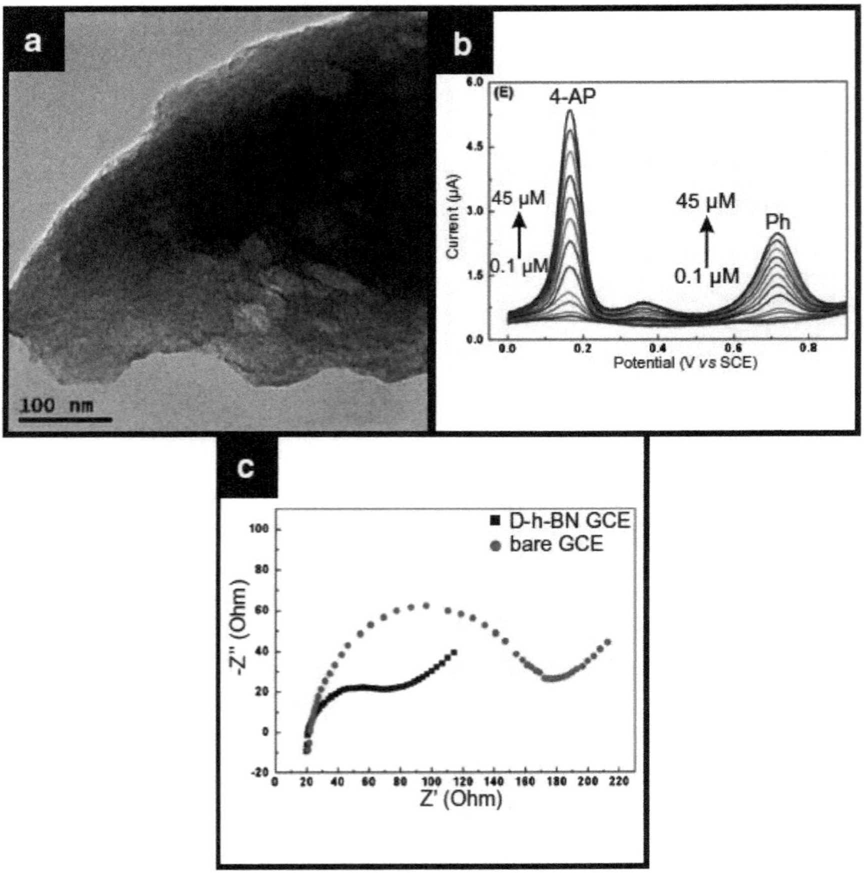

Figure 6.4 (a) TEM image of *h*-BN defect, (b) plot of 4-AP and phenol using D-*h*-BN modified GCE, and (c) EIS of D-*h*-BN modified GCE. (Reprinted from Ref. [66] with permission from Elsevier Copyright 2019 Elsevier.)

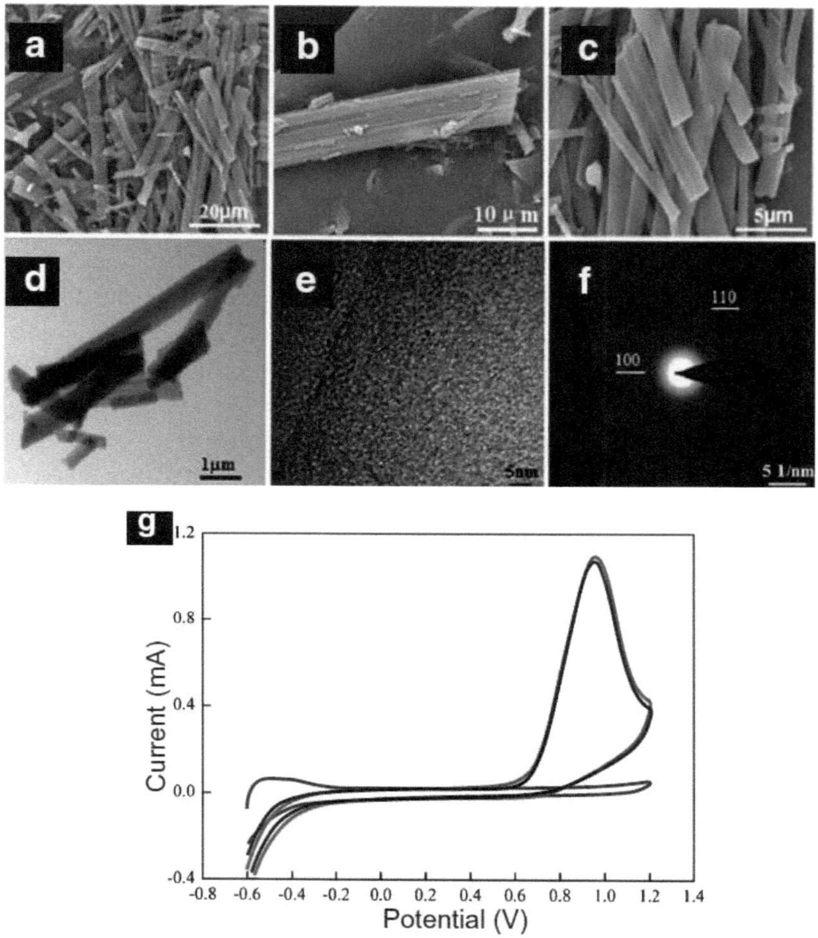

Figure 6.5 (a) Low magnification, (b, c) high-magnification SEM images, (d) TEM images, (e) HR-TEM, (f) SAED, and CV curves (g) of the crystallized (poor) *h*-BN whiskers (Red line), crystallized (highly) BN whiskers (Black line), and the unmodified Ti (Blue) for the electro-oxidation of nitrite (1.0 mM nitrate in 0.1 mM PBS). (Reprinted from Ref. [67] with permission from the Royal Society of Chemistry.)

GO and GO-based modified electrodes are exploited for electro-catalytic sensing and had noteworthy development in recent years. An electro-catalytic biosensor is an analytical tool and an electrical signal converted into a biological response by using electro-catalytic approaches that detect the concentrations of analytes and LOD, LR, and sensitivity [68]. From this viewpoint of electro-catalytic performance, the GO-based composites for electro-catalytic biosensing have the capability to deliver an appropriate micro-environment for biomolecule immobilization. Also, they enable electron transportation from the biomolecules to electrode substrates [69]. This kind of GO-based modified electrodes provide good LLOD, high sensitivity, instant response time, and improved peak-to-peak ratios, and they can be used for the development of most biosensors with industrial applications. Commonly, the combination of bioactive species such as metalloproteins and enzymes is fabricated on the surface of GO-based modified electrodes for electrochemical biosensors.

In current years, biological species electrodes based on GO-based composites have been involved in the electrochemical detection of analytes. For example, the detection of glucose plays a vital role in the analysis and treatment of diabetes. Literature reports have revealed that GO-based composites have a good unique performance toward the selection and sensitivity in terms of electrochemical determination of glucose [70–72]. Carboxylic and amine groups of the GO sheet for biodegradable GO-based glucose biosensors were prepared by Liu et al. [73], and they have wider LR, excellent reproducibility, good sensitivity, and stability. GOx/rGO/chitosan modified electrode was prepared and reported by Lin and co-workers [74] for the glucose sensor and it showed good LOD (0.02 mM), LR (0.08 to 12 mM), and sensitivity (37.93 $\mu A\ mM^{-1}\ cm^{-2}$) as shown in Fig. 6.6.

Figure 6.7 shows the FE-SEM pictures of PANI/GNRs prepared by a cost-effective technique and displayed excellent electro-catalytic detection toward methyl parathion (organophosphorus pesticide) [75–79]. Also, the authors did a real-time analysis of green beans, cabbage, and strawberry samples. The electron transportation that is enhanced between PANI and GNRs makes the nanocomposite extremely meet for electro-catalytic detection of pesticides. PANI-encapsulated GNRs nanocomposite was synthesized by in situ polymerization of aniline with GNRs [80].

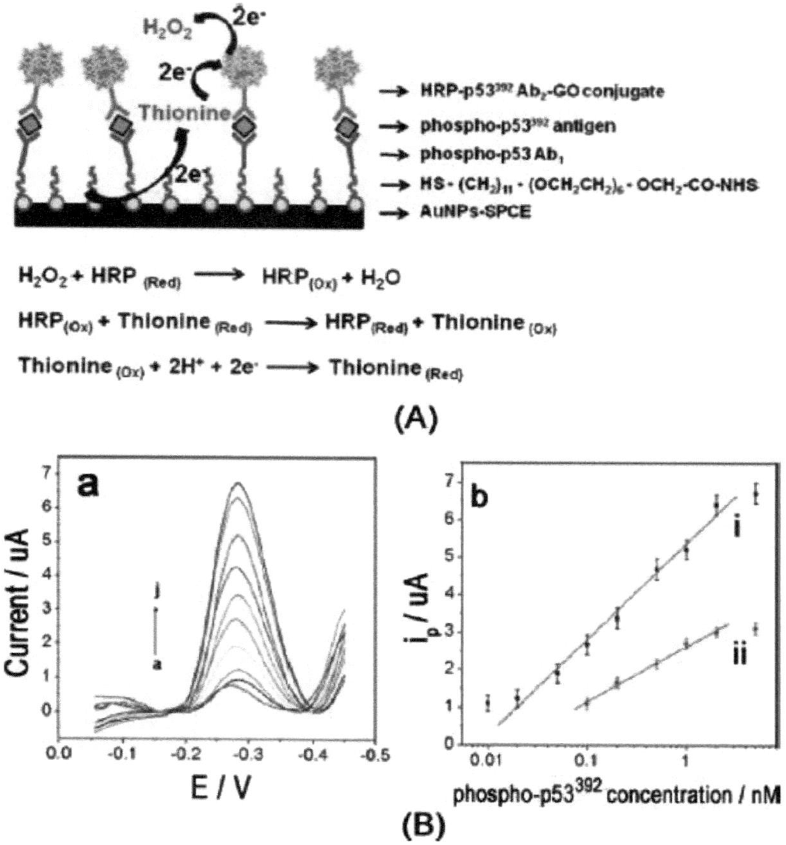

Figure 6.6 (A) Schematic illustration of the HRP–p53Ab$_2$–GO conjugate using multi-enzyme labeling amplification, (B) (a) SWV curves developed (after incubation) at HRP–p53Ab$_2$–GO/phospho-p53/Ab$_1$/NHS/AuNPs-SPCE, and (b) plots of the (i) modified HRP–p53Ab$_2$–GO/phospho-p53/Ab$_1$/NHS/AuNPs-SPCE and (ii) HRP–streptavidin –biotin–p53Ab$_2$/phospho-p53/Ab$_1$/NHS/AuNPs-SPCE for sensing phospho-p53. (Reprinted with permission from Ref. [72]. Copyright 2011 American Chemical Society.)

PtPd-doped reduced GONRs/GCE showed outstanding electrochemical sensing of TNT [81]. It showed good electrochemical sensing for TNT determination with LR of 0.01–3 ppm and an LOD of 0.8 ppb by using the ASV technique, as shown in Fig. 6.8.

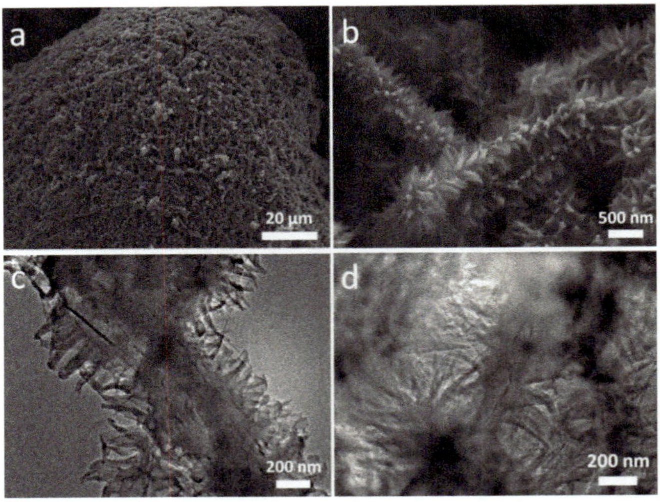

Figure 6.7 (a, b) SEM images, and (c, d) TEM images of PANI-GNRs composite. (Reproduced with permission from Reprinted from Ref. [80] with permission from American Chemical Society Copyright 2013).

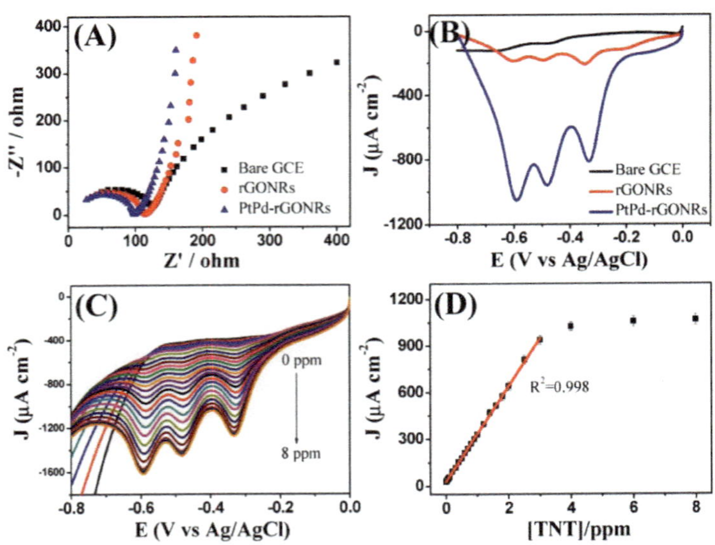

Figure 6.8 (A) EIS for bare GCE, rGONRs modified GCE, and PtPd-rGONRs/GCE in 5 mM $K_4[Fe(CN)_6]/K_3[Fe(CN)_6]$ containing 0.1 M KCl, (B) SV for 3 ppm of TNT attained at bare GCE, rGONRs-modified GCE, and PtPd-rGONRs-modified GCE in 0.1 M PBS (pH = 6.5) with 0.4 M KCl, (C) SV of PtPd-rGONRs-modified GCE with different concentrations of TNT in 0.1 M PBS (pH = 6.5) with 0.4 M KCl, and (D) plot between cathodic current and TNT concentration. (Reproduced with permission from Reprinted from Ref. [81] with permission from American Chemical Society Copyright 2015).

6.4 Conclusion

In this current review, we have discussed the current developments in synthesis and electrochemical sensor application. GO (graphene derivative) and *h*-BN showed significant importance in electro-catalytic studies and electro-analytical fields. Compared to other LMs, such as graphene oxide (GO), MXenes, hexagonal-boron nitride (*h*-BN), graphitic carbon nitride (g-C_3N_4), and molybdenum disulfide (MoS_2), in terms of electrochemical properties and applications include their synthesis technique, solubility, processability, conductivity, *h*-BN, and GO inexpensive source materials. Due to physical properties and structural morphologies, *h*-BN, GO, and their composites have been used to develop the modified electrodes for a wider range of applications in electrochemical sensors. These unique properties and advances in the sensor research field deliver novel sensitive and selective electrochemical detection of different biomarkers, toxic metal ions, food toxins, biological compounds (dopamine, ascorbic acid, etc.), and drugs that can deliver possible solutions not only in the electrochemical sensor and biosensor field but also in other field applications such as the water, agriculture, food industries, and the environmental toxic detection.

Acknowledgment

This work was supported by the Priority Research Program through the National Research Foundation of Korea (NRF) under the Ministry of Education (2018R1A6A1A03025526) and by the Basic Science Research Program through the NRF funded by the Ministry of Science, ICT (2021R1A2C1004540).

Abbreviations

AA	Ascorbic acid
AChE	Acetylcholinesterase
ASV	Adsorption stripping voltammetry
ClO_2	Chlorite
CV	Cyclic voltammetry
DA	Dopamine
GO	Graphene oxide

g-C_3N_4	Graphitic carbon nitride
GNR	Graphene nanoribbon
GOx	Glucose oxidase
GCE	Glassy carbon electrode
h-BN	Hexagonal boron nitride
HNO_3	Nitric acid
H_2O_2	Hydrogen peroxide
H_2SO_4	Sulfuric acid
H_3PO_4	Phosphoric acid
$KClO_3$	Potassium chlorate
$KMnO_4$	Potassium permanganate
LM	Layered material
LR	Linear range
LLOD	Lower limit of detection
MoS_2	Molybdenum disulfide
$NaNO_3$	Sodium nitrate
NT	Nanotube
NS	Nanosheet
PBS	Phosphate buffer solution
PANI	Polyaniline
Pb	Lead
QD	Quantum dot
RT	Room temperature
SPGME	Screen printed graphite macroelectrode
SV	Stripping voltammogram
TNT	Tri nitro toluene
UA	Uric acid
4-AP	4-Aminophenol

References

1. C. Tan, X. Cao, X.-J. Wu, Q. He, J. Yang, X. Zhang, J. Chen, W. Zhao, S. Han, G.-H. Nam, M. Sindoro, H. Zhang, *Chem. Rev.* 117, **2017**, 6225.
2. A. K. Geim, *Science* 324, **2009**, 1530.
3. X. Huang, X. Qi, F. Boey, H. Zhang, *Chem. Soc. Rev.* 41, **2012**, 666.
4. L. Li, Y. Yu, G. J. Ye, Q. Ge, X. Ou, H. Wu, D. Feng, X. H. Chen, Y. Zhang, *Nat. Nanotechnol.* 9, **2014**, 372.

5. Q. H. Wang, K. K. Zadeh, A. Kis, J. N. Coleman, M. S. Strano, *Nat. Nanotechnol.* 7, **2012**, 699.

6. K. F. Mak, C. Lee, J. Hone, J. Shan, T. F. Heinz, *Phys. Rev. Lett.* 105, **2010**, 136805.

7. K. K. Zadeh, J. Z. Ou, T. Daeneke, A. Mitchell, T. Sasaki, M. S. Fuhrer, *Appl. Mater. Today* 5, **2016**, 73.

8. H. Yu, R. Shi, Y. Zhao, T. Bian, Y. Zhao, C. Zhou, G. I. N. Waterhouse, L.-Z. Wu, C.-H. Tung, T. Zhang, *Adv. Mater.* 29, **2017**, 1605148.

9. L. Song, L. Ci, H. Lu, P. B. Sorokin, C. Jin, J. Ni, A. G. Kvashnin, D. G. Kvashnin, J. Lou, B. I. Yakobson, P. M. Ajayan, *Nano Lett.* 10, **2010**, 3209.

10. H. Zhang, *ACS Nano*, 9, **2015**, 9451–9469.

11. K. I. Bolotin, K. J. Sikes, Z. Jiang, M. Klima, G. Fudenberg, J. Hone, P. Kim, H. L. Stormer, *Solid State Commun.* 146, **2008**, 351.

12. Y. Ma, L. Kou, X. Li, Y. Dai, S. C. Smith, T. Heine, *Phys. Rev. B* 92, **2015**, 085427.

13. Y. Cao, V. Fatemi, S. Fang, K. Watanabe, T. Taniguchi, E. Kaxiras, P. J. Herrero, *Nature* 556, **2018**, 43.

14. M. F. El-Kady, V. Strong, S. Dubin, R. B. Kaner, *Science* 335, **2012**, 1326.

15. A. A. Balandin, S. Ghosh, W. Bao, I. Calizo, D. Teweldebrhan, F. Miao, C. N. Lau, *Nano Lett.* 8, **2008**, 902.

16. R. R. Nair, P. Blake, A. N. Grigorenko, K. S. Novoselov, T. J. Booth, T. Stauber, N. M. R. Peres, A. K. Geim, *Science* 320, **2008**, 1308.

17. B. L. Li, J. Wang, H. L. Zou, S. Garaj, C. T. Lim, J. Xie, N. B. Li, D. T. Leong, *Adv. Funct. Mater.* 26, **2016**, 7034.

18. K. F. Mak, J. Shan, *Nat. Photonics* 10, **2016**, 216.

19. D. Deng, K. S. Novoselov, Q. Fu, N. Zheng, Z. Tian, X. Bao, *Nat. Nanotechnol.* 11, **2016**, 218.

20. L. Feng, Z. Liu, *Nanomedicine* 6, **2011**, 317.

21. X. Hong, J. Liu, B. Zheng, X. Huang, X. Zhang, C. Tan, J. Chen, Z. Fan, H. Zhang, *Adv. Mater.* 26, **2014**, 6250.

22. Q. Yun, L. Li, Z. Hu, Q. Lu, B. Chen, H. Zhang, *Adv. Mater.* 32, **2020**, 1903826.

23. M. H. Rümmeli, C. G. Rocha, F. Ortmann, I. Ibrahim, H. Sevincli, F. Börrnert, J. Kunstmann, A. Bachmatiuk, M. Pötschke, M. Shiraishi, *Adv. Mater.* 23, **2011**, 4471.

24. G. Yang, L. Li, W. B. Lee, M. C. Ng, *Sci. Technol. Adv. Mater.* 19, **2018**, 613.
25. D. S. Shin, H. G. Kim, H. S. Ahn, H. Y. Jeong, Y.-J. Kim, D. Odkhuu, N. Tsogbadrakh, B. H. Kim, *RSC Adv.* 7, **2017**, 13979.
26. V. Kumar, A. Kumar, D.-J. Lee, S.-S. Park, *Materials* 14, **2021**, 4590.
27. C. K. Chua, M. Pumera, *Chem. Soc. Rev.* 43, **2014**, 291.
28. Z. Wang, J. Liu, W. Wang, H. Chen, Z. Liu, Q. Yu, H. Zeng, L. Sun, *Chem. Commun.* 49, **2013**, 10835.
29. X.-Y. Wang, A. Narita, K. Müllen, *Nat. Rev. Chem.* 2, **2017**, 100.
30. B. Brodie, *Ann. Chim. Phys.* 45, **1855**, 351.
31. B. C. Brodie, *Philos. Trans. R. Soc.* 149, **1859**, 249.
32. B. C. Brodie, *Philosophical Transactions The Royal Society,* 149, **1859**, 249.
33. L. Staudenmaier, *Ber. Dtsch. Chem. Ges.* 31, **1898**, 1481.
34. F. Pendolino, N. Armata, **2017**, DOI: doi.org/10.1007/978-3-319-60429-9.
35. N. I. Kovtyukhova, P. J. Ollivier, B. R. Martin, T. E. Mallouk, S. A. Chizhik, E. V. Buzaneva, A. D. Gorchinskiy, *Chem. Mater.* 11, **1999**, 771.
36. S. Eigler, M. E. Heim, S. Grimm, P. Hofmann, W. Kroener, A. Geworski, C. Dotzer, M. Röckert, J. Xiao, C. Papp, *Adv. Mater.* 25, **2013**, 3583.
37. H. Yu, B. Zhang, C. Bulin, R. Li, R. Xing, *Sci. Rep.* 6, **2016**, 36143.
38. D. C. Marcano, D. V. Kosynkin, J. M. Berlin, A. Sinitskii, Z. Sun, A. Slesarev, L. B. Alemany, W. Lu, J. M. Tour, *ACS Nano* 4, **2010**, 4806.
39. J. Zhang, Q. Liu, Y. Ruan, S. Lin, K. Wang, H. Lu, *Chem. Mater.* 30, **2018**, 1888.
40. X.-F. Jiang, Q. H. Weng, X.-B. Wang, X. Li, J. Zhang, D. Golberg, Y. Bando, *J. Mater. Sci. Technol.* 31, **2015**, 589.
41. J. Pu, K. Zhang, Z. Wang, C. Li, K. Zhu, Y. Yao, G. Hong, *Adv. Funct. Mater.* 31, **2021**, 2106315.
42. R. Han, F. Liu, X. F. Wang, M. H. Huang, W. X. Li, Y. Yamauchi, X. D. Sun, Z. G. Huang, *J. Mater. Chem. A* 8, **2020**, 14384.
43. R. Kumar, S. Sahoo, E. Joanni, R. K. Singh, R. M. Yadav, R. K. Verma, D. P. Singh, W. K. Tan, A. P. Pino, S. A. Moshkalev, A. Matsuda, *Nano Res.* 12, **2019**, 2655.
44. P. Sutter, J. Lahiri, P. Albrecht, E. Sutter, *ACS Nano.* 5, **2011**, 7303.
45. G. Y. Lu, T. R. Wu, Q. H. Yuan, H. S. Wang, H. M. Wang, F. Ding, X. M. Xie, M. H. Jiang, *Nat. Commun.* 6, **2015**, 6160.

46. F. Müller, K. Stöwe, H. Sachdev, *Chem. Mater.* 17, **2005**, 3464.

47. S. Chatterjee, Z. T. Luo, M. Acerce, D. M. Yates, A. T. C. Johnson, L. G. Sneddon, *Chem. Mater.* 23, **2011**, 4414.

48. J. Yu, L. Qin, Y. F. Hao, S. Y. Kuang, X. D. Bai, Y.-M. Chong, W. J. Zhang, E. G. Wang, *ACS Nano.* 4, **2010**, 414.

49. E. A. Pruss, G. L. Wood, W. J. Kroenke, R. T. Paine, *Chem. Mater.* 12, **2000**, 19.

50. S. Frueh, R. Kellett, C. Mallery, T. Molter, W. S. Willis, C. King'ondu, S. L. Suib, *Inorg. Chem.* 50, **2011**, 783.

51. K. H. Lee, H. J. Shin, J. Lee, I. Y. Lee, G. H. Kim, J. Y. Choi, S. W. Kim, *Nano Lett.* 12, **2012**, 714.

52. N. Guo, J. Wei, L. Fan, Y. Jia, D. Liang, H. Zhu, K. Wang, D. Wu, *Nanotechnology* 23, **2012**, 415605.

53. Y.-H. Lee, K.-K. Liu, A.-Y. Lu, C.-Y. Wu, C.-T. Lin, W. Zhang, C.-Y. Su, C.-L. Hsu, T.-W. Lin, K.-H. Wei, Y. Shi, L.-J. Li, *RSC Adv.* 2, **2012**, 111.

54. Y. Shi, C. Hamsen, X. Jia, K. K. Kim, A. Reina, M. Hofmann, A. L. Hsu, K. Zhang, H. Li, Z. Y. Juang, M. S. Dresselhaus, L. J. Li and J. Kong, *Nano Lett.* 10, **2010**, 4134.

55. Y. Gao, W. Ren, T. Ma, Z. Liu, Y. Zhang, W.-B. Liu, L.-P. Ma, X. Ma, H.-M. Cheng, *ACS Nano* 7, **2013**, 5199.

56. J.-H. Park, J. C. Park, S. J. Yun, H. Kim, D. H. Luong, S. M. Kim, S. H. Choi, W. Yang, J. Kong, K. K. Kim, Y. H. Lee, *ACS Nano* 8, **2014**, 8520.

57. V. Guerra, C. Wan, T. McNally, *Prog. Mater. Sci.* 100, **2019**, 170.

58. A. Vokhminstev, I. Weinstein, D. Zamyatin, *J. Lumin.* 208, **2019**, 363.

59. Y. Wu, P. Wu, Y. Chao, J. He, H. Li, L. Lu, *Nanotechnology* 29, **2017**, 025604.

60. A. Hayat, M. Sohail, M. S. Hamdy, T. A. Taha, H. S. AlSalem, A. M. Alenad, M. A. Amin, R. Shah, A. Palamanit, J. Khan, W. I. Nawawi, S. K. Baburao Mane, *Surf. Interfaces* 29, **2022**, 101725.

61. K. Uosaki, G. Elumalai, H. Noguchi, T. Masuda, A. Lyalin, A. Nakayama, *J. Am. Chem Soc.* 136, **2014**, 6542.

62. A. F. Khan, D. A. C. Brownson, E. P. Randviir, G. C. Smith, C. E. Banks, *Anal Chem.* 88, **2016**, 9729.

63. Q. Li, C. Huo, K. Yi, L. Zhou, L. Su, X. Hou, *Sens. Actuators B: Chem.* 260, **2018**, 346.

64. M. L. Yola, N. Atar, *Appl Surf Sci.* 458, **2018**, 648.

65. E. E. L. Tanner, R. G. Compton, *Electroanalysis* 30, **2018**, 1336.

66. T. R. Kiran, N. Atar, M. L. Yola, *J Electrochem Soc.* 166, **2019**, 495.
67. M. L. Yola, N. Atar, *Mater Sci Eng C.* 69, **2019**, 669.
68. R. Jerome, A. K. Sundramoorthy, *Anal Chim Acta.* 1132, **2020**, 110.
69. T. Kokulnathan, R. Vishnuraj, T.-J. Wang, E. A. Kumar, B. Pullithadathil, *Ecotoxicol Environ Saf.* 207, **2021**, 111276.
70. Y. Shen, H. Yan, H. Guo, Y. Long, W. Li, *Sens. Actuators B Chem.* 303, **2020**, 127248.
71. Y. Shen, H. Ouyang, W. Li, Y. Long, *Microchimica Acta* 188, **2020**, 40.
72. W. Luo, T. Yang, L. Su, K.-C. Chou, X. Hou, *RSC Adv.* 6, **2016**, 27767.
73. E. Bakker, M. T. Diaz, *Anal. Chem.* 74, **2002**, 2781.
74. Z. J. Wang, X. Z. Zhou, J. Zhang, F. Boey, H. Zhang, *J. Phys. Chem. C* 113, **2009**, 14071.
75. T. T. Baby, S. S. J. Aravind, T. Arockiadoss, R. B. Rakhi, S. Ramaprabhu, *Sens. Actuators, B* 145, **2010**, 71.
76. D. Du, L. M. Wang, Y. Y. Shao, J. Wang, M. H. Engelhard, Y. H. Lin, *Anal. Chem.* 83, **2011**, 746.
77. Y. Liu, D. S. Yu, C. Zeng, Z. C. Miao, L. M. Dai, *Langmuir* 26, **2010**, 6158.
78. X. H. Kang, J. Wang, H. Wu, I. A. Aksay, J. Liu, Y. H. Lin, *Biosens. Bioelectron.* 25, **2009**, 901.
79. M. Govindasamy, V. Mani, S.-M. Chen, T.-W. Chen, A. K. Sundramoorthy, *Sci. Rep.* 7, **2017**, 46471.
80. L. Li, A. R. O. Raji, H. Fei, Y. Yang, E. L. Samuel, J. M. Tour, *ACS Appl. Mater. Interfaces* 5, **2013**, 6622.
81. R. Zhang, C.-L. Sun, Y.-J. Lu, W. Chen, *Anal. Chem.* 87, **2015**, 12262.

Chapter 7

Influences of Graphene, Transition Metal Dichalcogenides, and MXenes Nanocomposites as Energy Storage Devices

Kalimuthu Kanimozhi[a] **and Vengatesan M. Rengaraj**[b]

[a]*Department of Chemistry, University College of Engineering, BIT Campus, Tiruchirappalli – 620 024, India*
[b]*Department of Chemical Engineering, Khalifa University, Abu Dhabi, UAE*

kanirams47@gmail.com

The 2D nanomaterials are being investigated more interestingly due to their outstanding properties, including the electrical, physicochemical, optical, micromechanical, and thermoelectric characteristics. 2D materials are considered superconductive materials, which exhibit extraordinary behavior compared to their bulk due to a direct bandgap, larger specific surface area, strong covalent bonds on the nanosheet, and the weak van der Waals interactions between the 2D layers. These characteristics result in a large gap between the exfoliated 2D nanosheets, thus providing an enormous number of active sites and increasing

Layered Materials: Synthesis and Applications in Energy and Environmental Remediation
Edited by Alagarsamy Pandikumar, Kandasamy Jothivenkatachalam, and Subramanian Moscow

ISBN 978-981-5129-17-5 (Hardcover), 978-1-003-50831-1 (eBook)
www.jennystanford.com

the accessibility of the material's surface. Consequently, 2D nanomaterials are exploited ever more in several applications such as energy conversion/storage, hydrogen production, and gas storage. This book chapter covers the recent research of thin film of 2D materials using various novel technologies to synthesize and grow monolayer 2D materials on different substrates in different fields based on the knowledge available in the literature. In this chapter, we will discuss the properties and processing of graphene, transition metal dichalcogenides, MXene nanocomposites, and their applications in the field of energy storage materials.

7.1 Introduction

The strong demand for futuristic energy-storage materials and devices is exceptionally increasing, owing to the request for more powerful energy-storage systems with excellent power density and better cycle lifetime [1, 2]. For this reason, serious efforts have been undertaken to improve the electrode performance to achieve significantly improved capacity, high-rate capability, and longer cycle life [3, 4]. One of the growing applications is to use these 2D materials as potential electrodes for rechargeable batteries and electrochemical capacitors. The 2D materials are a new type of sheet-like nanomaterials, typically showing lateral size from hundreds of nanometers to a few micrometers and their average thickness is <5 nm [5]. The study of 2D materials was initiated after the preparation of the monolayer graphene sheet from graphite in 2004. These materials have a large surface area with a high percentage of atoms exposed to the surface, which makes them have a good interaction with the matrix and even a small amount of them can greatly increase the performance of polymers. In addition, their abundant species make them suitable fillers for different application purposes [6, 7]. Graphene and its analogous materials such as the graphitic carbon nitride (g-C_3N_4), the transition metal dichalcogenides (TMDs), black phosphorus (BP), silicene, hexagonal boron nitride (h-BN) MXenes, and layered double hydroxides (LDHs) are considered as fascinating materials, owing to their outstanding electronic, physical, optical, chemical, and mechanical properties.

It has been reported that with a small loading (<1–5 wt%) of 2D materials, such as layered silicates or graphene, into a polymer matrix, the mechanical properties of the resulting polymer composites improved up to ~200% compared to the neat polymers [8–12]. In this book chapter, we focus on the recent progress of 2D transition metal dichalcogenides, MXenes, and graphene materials-based polymer composites with their applications in energy storage aspects.

7.2 Graphene

The graphene is composed of carbon atoms tightly packed into a 2D honeycomb lattice structure through sp2 hybridization. The spongy structure in graphene makes it ideal for reinforcement of polymer matrices. It implies that matrices for graphene provide high strength and increase hydrophobicity as well as high conductivity. Such a structure displays versatile physical and chemical properties, such as high carrier mobility at room temperature (~10,000 $cm^2\ V^{-1}\ s^{-1}$), high Young's modulus (~1 TPa), large theoretical specific surface area (2630 $m^2\ g^{-1}$), low light absorption (~2.3%), and high thermal conductivity (~5000 $Wm^{-1}\ K^{-1}$).

However, 2D graphene nanosheets have a unique band structure, in which the electrons can freely delocalize at a faster speed than the speed of light. The conductivity as well as charge carrier mobility can be utilized to accumulate and store charge, which is the fundamental mechanism of energy storage in capacitors. GO is an essential derivative of graphene, which was obtained by strong oxidation of the graphite with oxidative agents.

During the oxidation process, various oxygen-containing functional groups such as carboxyl, hydroxyl, and epoxy groups are introduced on the surface of graphene. Due to graphene's vast mobility of charge carrier property, which is capable of passing thousands of interatomic paths without scattering, this material has strong potential to be utilized for any electronic device [13, 14].

For example, GO has a much poorer electrical conductivity than graphene, due to the transformation of C–C bonds from sp^2 to sp^3 hybridization by the strong oxidation process. The graphene is insoluble in water, and GO exhibits excellent dispersity in water due to the large amount of hydrophilic oxygen-containing groups on the surface. Therefore, graphene and GO show different application purposes for polymer composites. The traditional lithium-ion battery could soon be replaced in electric cars by a new generation made from graphene. This material has many qualities. In particular, it should allow the battery to be recharged much more rapidly than today. Thinner devices or larger capacities without requiring extra room. Not only that, but graphene allows for much higher capacities.

Synthesis of graphene refers to any process for fabricating graphene, depending on the desired size, purity, and crystallinity of the final product. Common graphene synthesis approaches include mechanical exfoliation from graphite, chemical vapor deposition, and reduction of graphene oxide through heating (Fig. 7.1).

7.2.1 Synthesis of Graphene

The synthesis of monolayer graphite was tried in 1975, by the formation of mono- and multi-layered graphite by thermal decomposition of carbon on single crystal Pt substrates. However, due to a lack of consistency between the properties of such sheets, formed on different crystal planes of Pt, and failure to identify the beneficial applications of the product, the process was not studied extensively, at that period of time. The discovery of graphene has been credited since 1975 with various treatments. They have first shown repeatable synthesis of graphene through exfoliation. The technique has been and is being followed since then, along with efforts to develop new processing routes for efficient synthesis of large-scale graphene [15, 16]. The synthesis routes of graphene can be broadly categorized into different sections, as detailed below.

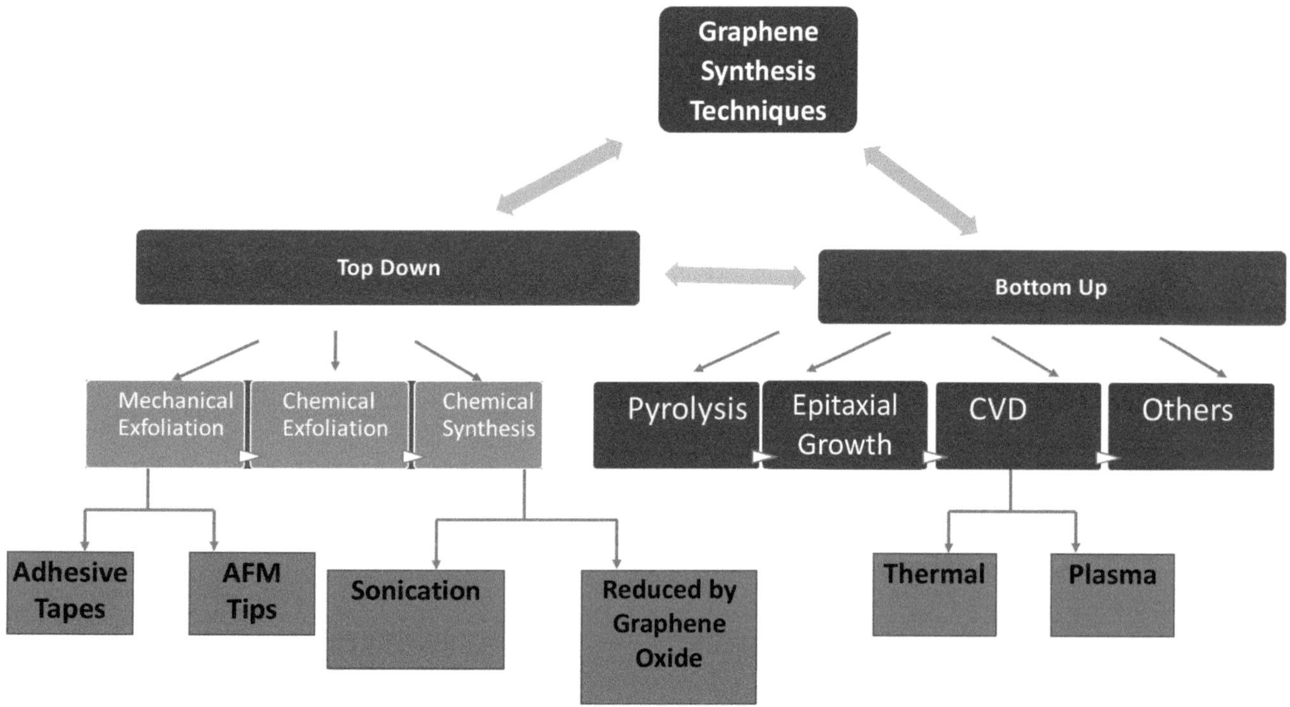

Figure 7.1 Graphene synthesis techniques.

7.2.2 Exfoliation and Cleavage

Graphite is stacked layers of many graphene sheets, bonded together by weak van der Waals force. Thus, in principle, it is possible to produce graphene from a high-purity graphite sheet, if the bonds are broken. The exfoliation and cleavage use mechanical or chemical energy to break these weak bonds and separate individual graphene sheets. The intercalation of pure graphite sheet from potassium metal and then exfoliating with ethanol to form dispersed C sheets. During sonication, the exfoliated nano-carbon sheets formed nanoscrolls. TEM analysis showed the presence of 40 ± 15 layers in each sheet. The positive side of the approach includes separated layers of graphene from graphite, carbon nanoscrolls were much thicker than a few-layer graphene, and the process was in practice too. The process needed modification to finally produce FLG. It took almost 20 more months to find a solution, reported FLG, and even SLG formation by exfoliation.

The exfoliation is basically a repeated peeling process. In the study of a commercially available highly oriented pyrolytic graphite (HOPG) sheet of 1 mm thickness was subjected to dry etching in oxygen plasma to make many 5 μm deep mesas (of area 0.4–4 mm^2). This was then put on a photoresist and baked, to stick the mesas to the photoresist. Then, scotch tape was used to peel off layers from the graphite sheet. Thin flakes, attached to the photoresist, released in acetone and transferred to a Si substrate, were found to have from single to few-layer graphene sheet [17–19]. The technique was later used to produce 2D atomic crystals of many other materials, including BN and MoS_2. In slight variations of the original process, it was shown that large (~10 μm) and flat graphene flakes can be produced by manipulating the substrate bonding of HOPG on Si substrate and controlled exfoliation. In another approach, mm-sized single to few-layer graphene was produced by bonding bulk graphite to borosilicate glass followed by exfoliation, to leave single or few layers of graphene on the substrate. Both of these advancements stressed on modifying the bonding with the substrate to generate large-area graphene sheets.

In a little different approach of exfoliation in the liquid phase, with the use of hydrophobicity of graphite oxide (GO) and exfoliated GO nanosheets by ultrasonication in aqueous suspension and attempted reduction of the films in hydrazine hydrate at 100 °C for 24 h. The product, however, was not fully reduced and was found to have some amount of oxygen left in the structure. Hence, it can be called *partially reduced exfoliated nano graphite oxide sheets.* Acting on the pros and cons of this liquid phase exfoliation process, there was a new way of producing single- to few-layer graphene sheets, through dispersion and exfoliation of pure graphite in N-methyl-pyrrolidone. The monolayer yield was 1 wt%, which could be improved up to 12 wt%, with further processing.

The energy required to exfoliate graphite into single-layer graphene was countered by the solvent-graphene interaction, with the solvent having similar surface energy as that of graphene. This process of liquid-based exfoliation of graphite in organic solvent has the promise of large-scale production of graphene. A similar kind of process was used to produce single- to few-layer graphene, by making dispersion of graphite powder in sodium dodecylbenzene sulfonate (SDBS), followed by sonication, to exfoliate the graphite into graphene [20, 21]. There were many such attempts to produce graphene either from graphite or graphite oxide powder and using different solvents. However, the liquid phase exfoliated products suffer from the limitation that the structure has a lot of defects, due to oxidation and reduction processes, leading to much poor electrical properties of graphene. Future improvisations need to concentrate on controlling the number of layers and minimizing impurity levels to lead this process to an industrial scale-production level.

7.2.3 Graphene Pyrolysis

The manufacturing of graphene can also be done via chemical synthesis in the bottom-up process. Thermal reaction is done using an equimolar ratio of ethanol and sodium in an autoclave. Pyrolization of sodium ethoxide via sonication can be used to detach graphene smoothly. The process leads to the production of graphene, which is 10 μm. The crystalline structure, varying layers, and graphitic nature were inveterated via SAED, TEM, and Raman spectroscopy. The results obtained from the Raman spectroscopy showed a broad D – band and G – band. The intensity of the ratio of IG/ID was approximately 1.6 and this showed a defective graphene. The advantage of this approach is enormous. The pyrolysis of graphene through biomass precursor material is shown in Fig. 7.2.

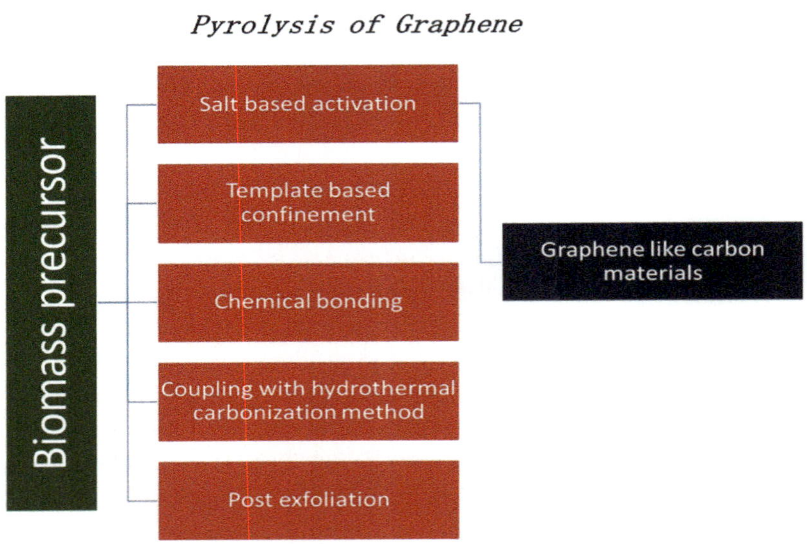

Figure 7.2 Pyrolysis of graphene from biomass precursor.

Some of these merits include low cost and easily fabricated for high purity. In spite of these advantages, some limitations of using this approach include a large number of defects being produced [22].

7.2.4 Chemical Vapor Deposition (CVD)

A process where molecules are heated is described as CVD. The substrate is first diffused onto a thermally disintegrated precursor at elevated temperatures. The deposition is usually done on a thin film, crystalline, solid, liquid, or gaseous precursor. This process is usually carried out on Cu, Ni, Pd, Ru, and Ir metals. The CVD growth of graphene was tested first on nickel. Ni was introduced to methane with a temperature around 900 °C, which resulted in the formation of thin graphite. Investigation into the evaporation of carbon from graphite rods was also carried out. In 1975, mono-layer graphitic materials were deposited on platinum. Other researchers also investigated graphene grown on metal surfaces (Ir) in 1984. The information compiled from the literature shows that several studies have been conducted in terms of the physical and chemical characteristics of graphene, hence creating a new research direction (graphene-based electronics). The possibility of graphene synthesis on Ni foil with the aid of CVD using camphor was carried out in 2006 [23].

7.3 Energy Storage/Conversion Devices

Graphene is ideal for the production of energy storage devices. When it is treated with metal oxide, it shows limited sheet restacking. Lithium-ion stores up to 180 Wh of energy per kilogram while graphene can store up to 1,000 Wh kg^{-1}. The interconnected networks of graphene being highly conductive contribute to its interest in energy storage applications. The merits of graphene such as stable electrochemical, good mechanical stability, and porous microstructure, make use of it as an excellent energy storage device. These energy storage materials are used as fuel cells, solar cells, and batteries, as well as supercapacitors. A monolayer of graphene sufficient in size to cover an entire soccer field would weigh less than 3 g (graphene's surface area is 2630 square meters per gram). This huge surface area associated with this small amount of graphene can be squeezed inside an AA battery, enabling battery designs with the ability to store massive amounts of charge. However, the full potential of energy-storage devices built from graphene has yet to be realized. Many challenges remain, particularly with regard to feasible techniques for the low-cost mass production of graphene with controlled microstructure and low residual oxygen content. The supports in proton exchange membrane (PEM) fuel cells are often made up of graphene while for batteries like Li batteries, they are utilized in the cell as anodic and cathodic material. In supercapacitors, graphene is used as the electrode material for both the double-layer capacitors as well as the pseudo capacitors. They are also utilized in fuel cells as dye sensitizers [16, 23].

7.4 Transition Metal Dichalcogenides (TMDs)

2D TMDs have a layered structure as another bulk source of 2D nanomaterials such as graphite, which display strong potential to be exploited in several applications due to their outstanding properties. Some of the TMDs have a semiconducting nature with a tunable bandgap. In comparison with graphene, which can be only activated through functionalization using favorable compounds or molecules, TMDs are chemically adaptable. The versatile TMDs have a crystalline structure with the formulae of MX_2. In this formula, M stands for transition metal (e.g., Ti, Ta, Mo, W, V, Re), and X stands for an element that is a chalcogenide component (e.g., S, Te, Se).

Any single layer of TMD includes three atomic sheets linked together through the van der Waals interactions. The transition metal is placed amongst the two chalcogen layers. Several varieties of the TDMs class of 2D materials such as ZrS_2, MoS_2, WS_2, $MoSe_2$, ReS_2, and WSe_2 are exploited for essential and industrial research as in energy storage, catalysis, and electronic and sensing devices. The bulk-sized TMDs are varied from being insulators (e.g., HfS_2), semiconductors (e.g., WS_2, MoS_2), semimetals (e.g., $TiSe_2$, WTe_2), and exactly metals (e.g., VSe_2, NbS_2). Furthermore, a few examples of the bulk-sized TDMs (e.g., TaS_2, $NbSe_2$) display significant properties in lower temperatures such as superior conductivity, metal- to non-metal transition (Mott transition phenomenon), and charge density wave (CDW), which is defined as a distortion of the crystal lattice periodically. Similar to other 2D nanomaterials, the physicochemical properties of TDMs could be improved through the exfoliation process, which leads to exhibiting supplementary features as a result of the confinement effect. Facile synthesis, high mechanical and chemical stability, reasonable cost, and high specific surface area, make TDMs suitable for vast fields of applications [24–26].

7.5 MXene Materials

MXene is another class of 2D materials (2D transition metal carbides, carbonitrides, and nitrides), where the 2D nanoflake of MXene is comprised of $n + 1$ ($n = 1, 2, 3$) layers (M layers) of early transition metals (e.g., V, Tr, Cr, Mo, W) that are inserted with n sheets of carbon element or nitrogen atoms (X elements). MXene can be represented through the formula of $M_{n+1}X_nT_x$, in which T_x defines the surface termination atoms (e.g., OH, O, F, Cl) that are linked to the external M layers. Due to the variety of MXene materials' structures and the vast selection of their elements, this family of 2D materials is growing rapidly. In contrast to graphene, MXene material cannot be delaminated from the bulk size material. MXene can be synthesized through the epitaxially grown process on the substrate by various etching and delamination processes.

Different MAX phases, such as Ti_2AlC material, can be exploited as a precursor to fabricate the 2D MXene. Similar to the MAX phases, MXene materials without surface terminations exhibited significant electronic properties, making them a strong candidate for electronic applications, such as metallic electrical conductors. The electronic characteristics of MXene materials can be tuned by altering the surface termination nature and changing the M groups; thus, they can behave as metal, semiconductor, or insulator material from metallic to semiconducting or even topological insulators. Owing to the high conductivity of MXenes and their layered structure, which eases ion diffusion, these materials are considered promising in energy storage applications such as electrode materials in batteries and supercapacitor devices [27–29].

7.6 Applications of the 2D Materials/Polymer Composites in Energy Storage

A fundamental understanding of the structures of electrodes, the electrode/electrolyte interfaces, and charge storage mechanisms play a key role in layered material applications in energy storage.

The demand for the storage of the conversed energies from solar and wind power with high efficiency has increased, to improve both the power density and energy density of the devices. To solve this problem, storing energy in electrochemical devices is one of the main strategies. There are two main reasons: (i) The restacked 2D materials provide a large interlayer spacing, making it easy for the insertion and extraction of ions, such as the Li^+, K^+, and OH^-. (ii) There are more opportunities for the ions binding to the active sites of the 2D materials due to their high theoretical specific surface area.

Additionally, the effect of defects on the electrochemical properties of layered materials needs to be studied. The overall electrochemical performance (e.g., gravimetric energy density and power density) will improve by increasing the operating voltage and choosing an appropriate electrolyte with a high operating voltage window (organic electrolytes or ionic liquids).

For example, graphene is one of the most popular 2D materials, showing a unique structure with high specific surface area and electrical conductivity. Therefore, graphene and its derivatives perform an important role in the application of energy-storage systems. In the following section, we will discuss the use of polymer-based composites in popular energy-storage systems, such as lithium-sulfur batteries (LSBs), and supercapacitors. First, we will review the 2D-based polymer composites as separators for battery applications. Then, the application of supercapacitors with 2D-based polymer composites acting as electrode materials will be summarized. For each aspect, the main development in 2D materials history and the challenges will be discussed [30–33].

7.7 Li-S Battery Separators

LIBs are widely used in our daily lives, such as cell phones, laptops, and electric cars, due to their high energy density, rechargeability, and safety. In the commercially used LIBs, $LiCoO_2$, graphite, and polymer membrane either PVDF or PP have been used as a cathode, anode, and separator, respectively. The energy density for this LIB was 150–190 Wh kg^{-1}, while the result could be obtained up to ~240 Wh kg^{-1} for commercial LIBs. In recent years, the idea of using lithium batteries for vehicle applications has received much research attention because this type of battery system in vehicles will reduce the use of fossil fuels and consequently air pollution will also be decreased. However, for vehicle applications, higher energy density (600–1300 Wh kg^{-1}) should be provided. Therefore, many advanced battery systems have been proposed and studied, such as Li-air batteries and Li-S batteries.

As one of the most promising systems, the Li-S battery was calculated to have a theoretical energy density of ~2600 Wh kg^{-1}, which was 10-fold that of the $LiCoO_2$ cathode battery. However, the shuttle effect of sulfur is one fatal restriction for the real use of Li-S batteries. The formed polysulfide (Li_2S) during cycling would be dissolved in the electrolyte and would go through the separator, leading to a sharp decrease in capacity and poor performance of the cycle life of the Li-S cells. More severely, the Li-S battery will "die" [34, 35]. Therefore, it is necessary to use a suitable separator to prevent the transfer of the polysulfide and ensure the normal diffusion of other important ions. Although porous polymer membranes are widely used for LIB separators such as PVDF and PP, these are not suitable candidates for the LSB system. For example, PP is considered to be a poor membrane material due to its large pore size (~100 nm), which is not adequate for preventing the transportation of polysulfide. To overcome this problem, direct coating graphene on the surface of the PP separator and the polymer composite membrane played a role in the current collector. The area on the separator coated by graphene provided a useful place to store the polysulfide dissolved in the electrolyte, retarding polysulfide diffusion and therefore increasing the cycle life of Li-S cells. Zhuang et al. reported a novel ternary-layered structure of the separator by integrating Nafion, PP, and GO together. The reason for using GO was that there were many functional groups on the surface of GO, such as carboxyl and hydroxyl groups, which exhibited a strong repulsion to the polysulfide anions due to the electrostatic interaction. The decay of capacitance showed a significant reduction to 0.18% per cycle, while the reduction per cycle for pure PP separator was 0.34% [36–38].

7.8 Supercapacitor Electrodes

The supercapacitor is another type of novel energy storage device with a higher power density (~1000 W kg^{-1}) and excellent cycling stability but lower energy density than traditional batteries. Based on large power density, supercapacitors have attracted much attention in various commercial applications, such as transportation and elevators. Supercapacitors can be divided into two categories based on different charge storage mechanisms, i.e., electric double-layer capacitor (EDLC), where the charge is absorbed on the surface, and pseudo capacitors with redox reaction occur on the surface to store charges (Fig. 7.3). Generally, both mechanisms will contribute to the capacitance of the electrode material. The strong π–π interaction between graphene layers makes them easily restacked, leading to the poor performance of the electrode. All these experiments demonstrated the important role of 2D materials in improving the electrical conductivity and capacitance of the polymer matrix [39, 40].

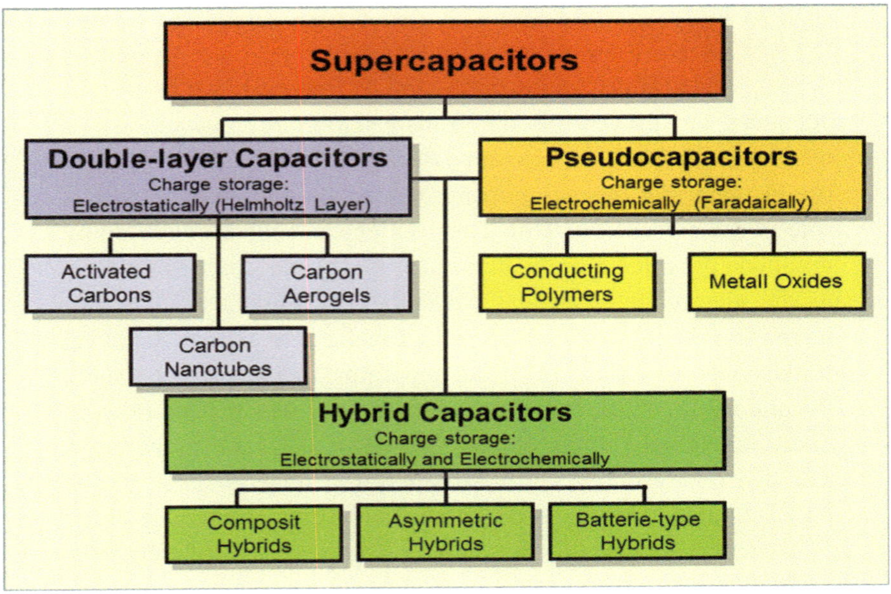

Figure 7.3 Supercapacitors.

7.9 Summary and Conclusion

The applications of 2D materials-based polymer composites such as graphene, MXenes, and transchalconides that are applied in Li-S batteries and supercapacitors have been summarized. The conducting 2D-based polymer composites provide a large surface area with many active sites, leading to an excellent performance for the electrochemical reaction on the supercapacitor. For battery separator applications, both 2D materials and polymers have porous structures, which prevent the transportation of other ions except Li^+ through separators, increasing the Coulombic efficiency of the batteries. Although the enormous potential of 2D materials-based polymer composites has been demonstrated in various fields, there are still many challenges in their practical applications. The high quality of graphene can be obtained by the scotch tape method; however, the extremely low quantity limits its practical use. Processing and applications of 2D layered nanostructures have developed, to meet the several challenges in this field. Moreover, methods have a major influence on the performance, reproducibility, and reliability of 2D nanomaterial heterostructure applications. The synthesis of 2D nanomaterials is important; therefore, careful consideration will be needed. There are also numerous important challenges to develop for commercialization.

References

1. Pushpendra Kumar et al., Review—Two-dimensional layered materials for energy storage applications, *ECS J. Solid State Sci. Technol.*, 5 Q3021, 2016.
2. P. G. Bruce, S. A. Freunberger, L. J. Hardwick, and J. M. Tarascon, *Nat Mater*, 11, 19, 2012.
3. P. Kumar, F. Y. Wu, L. H. Hu, S. A. Abbas, J. Ming, C. N. Lin, J. Fang, C. W. Chu, and L. J. Li, *Nanoscale*, 7, 8093, 2015.
4. J. Ming, M. Li, P. Kumar, and L. J. Li, *Acs Nano*, 10, 6037, 2016.
5. E. J. Markvicka, M. D. Bartlett, X. Huang, and C. Majidi, An autonomously electrically self-healing liquid metal–elastomer composite for robust soft-matter robotics and electronics, *Nature Materials*, 17 (7), 618–624, 2018.

6. C. Tan, X. Cao, X.-J. Wu et al., Recent advances in ultrathin two-dimensional nanomaterials, *Chemical Reviews*, 117 (9), 6225–6331, 2017.
7. C. Huo, Z. Yan, X. Song, and H. Zeng, 2D materials via liquid exfoliation: A review on fabrication and applications, *Science Bulletin*, 2015, 60, 1994.
8. A. K. Geim, Graphene: Status and prospects, *Science*, 324 (5934), 1530–1534, 2009.
9. G. Eda, Y. Y. Lin, C. Mattevi et al., Blue photoluminescence from chemically derived graphene oxide, *Advanced Materials*, 22 (4), 2392, 2010.
10. C. Zhi, Y. Bando, C. Tang, H. Kuwahara, and D. Golberg, Large-scale fabrication of boron nitride nanosheets and their utilization in polymeric composites with improved thermal and mechanical properties, *Advanced Materials*, 21 (28), 2889–2893, 2009.
11. Q. H. Wang, K. Kalantar-Zadeh, A. Kis, J. N. Coleman, and M. S. Strano, Electronics and optoelectronics of two-dimensional transition metal dichalcogenides, *Nature Nanotechnology*, 7 (11), 699–712, 2012.
12. T. Rodenas, I. Luz, G. Prieto et al., Metal–organic framework nanosheets in polymer composite materials for gas separation, *Nature Materials*, 14 (1), 48–55, 2014.
13. Suresh C. Pillai and Priyanka Ganguly, *IOP Publishing 2D Materials for Energy Storage and Conversion*, 2021, ISBN: 978-0-7503-3317-7
14. J. Ortiz Balbuena, P. Tutor De Ureta, E. Rivera Ruiz, and S. Mellor Pita, *Enfermedad de Vogt-Koyanagi-Harada Med. Clin. (Barc)*, 2016.
15. Wonbong Choi Indranil Lahiri, Raghunandan Seelaboyina, and Yong Soo Kang, Synthesis of graphene and its applications: A review, *Critical Reviews in Solid State and Materials Sciences*, 35 (1), 52–71, DOI: 10.1080/10408430903505036.
16. A.G. Olabi, Mohammad Ali Abdelkareem, Tabbi Wilberforce, and EnasTaha Sayed, Application of graphene in energy storage device: A review, *Renewable and Sustainable Energy Reviews*, 135, 110026, 2021.
17. M. Choucair, P. Thordarson, J. A. Stride, Gram-scale production of graphene based on solvothermal synthesis and sonication, *Nat Nanotechnol.*, 4 (30), 2009.
18. M. Chen, T. Tao, L. Zhang, W. Gao, and C. Li, Highly conductive and stretchable polymer composites based on graphene/MWCNT network, *ChemCommun*, 49, 1612–4, 2013.

19. D. Wei, Y. Liu, Y. Wang, H. Zhang, L. Huang, and G. Yu, Synthesis of N-doped graphene by chemical vapor deposition and its electrical properties, *Nano Lett*, 9, 1752–8, 2009.

20. Q. Yu, J. Lian, S. Siriponglert, H. Li, Y. P. Chen, and S-S Pei, Graphene segregated on Ni surfaces and transferred to insulators, *Appl Phys Lett*, 93, 113103, 2008.

21. A. Obraztsov, E. Obraztsova, A. Tyurnina, and A. Zolotukhin, Chemical vapor deposition of thin graphite films of nanometer thickness, *Carbon*, 45, 2017–21, 2007.

22. N. Gall, E. Rut'Kov, and A. Y. Tontegode, Intercalation of nickel atoms under two-dimensional graphene film on (111) Ir, *Carbon*, 38, 663–7, 2000.

23. M. A. Worsley, P. J. Pauzauskie, T. Y. Olson, J. Biener, J. H. Satcher Jr, and T. F. Baumann, Synthesis of graphene aerogel with high electrical conductivity, *J Am Chem Soc.*, 132, 14067–9, 2010.

24. L. M. Xie, Two-dimensional transition metal dichalcogenide alloys: Treparation, characterization and applications, *Nanoscale*, 7, 18392–401, 2015.

25. M. Chhowalla, H. S. Shin, G. Eda, L. J. Li, K. P. Loh, and H. Zhang. The chemistry of two dimensional layered transition metal dichalcogenidenanosheets, *Nat. Chem.*, 5, 2013.

26. Y. Gogotsi and B. Anasori, The rise of MXenes, *ACS Nano*, 13, 8491–4, 2019.

27. B. Anasori et al., Two-dimensional, ordered, double transition metals carbides (MXenes), *ACS Nano*, 9, 9507–16, 2015.

28. L. Verger, V. Natu, M. Carey, and M. W. Barsoum, MXenes: An introduction of their synthesis, select properties, and applications, *Trends Chem.*, 1, 656–69, 2019.

29. Wei Liu, Bakhtar Ullah, Ching-Ching Kuo, and Xingke Cai, Two-dimensional nanomaterials-based polymer composites: fabrication and energy storage applications, *Advances in Polymer Technology*, 4294306, 15, 2019.

30. A. B. Yaroslavtsev, T. Kulova, and A. M. Skundin, Electrode nanomaterials for lithium-ion batteries, *Russian Chemical Reviews*, 84 (8), 826, 2015.

31. P. G. Bruce, S. A. Freunberger, L. J. Hardwick, and J.-M. Tarascon, Li–O_2 and Li–S batteries with high energy storage, *Nature Materials*, 11 (1), 19, 2011.

32. A. Khaligh and Z. Li, Battery, ultracapacitor, fuel cell, and hybrid energy storage systems for electric, hybrid electric, fuel cell, and plug-in hybrid electric vehicles: State of the art, *IEEE transactions on Vehicular Technology*, 59 (6), 2806–2814, 2010.

33. R. Cao, W. Xu, D. Lv, J. Xiao, and J.-G. Zhang, Anodes for rechargeable lithium-sulfur batteries, *Advanced Energy Materials*, 5 (16), 1402273, 2015.

34. G. Zhou, L. Li, D.-W. Wang et al., A flexible sulfur-graphene polypropylene separator integrated electrode for advanced Li-S batteries, *Advanced Materials*, 27 (4), 641–617, 2015.

35. T. Z. Zhuang, J. Q. Huang, H. J. Peng et al., Rational integration of polypropylene/graphene oxide/nafion as ternary-layered separator to retard the shuttle of polysulfides for lithium-sulfur batteries, *Small*, 12 (3), 381–389, 2016.

36. Zahra Sayyar and Zahra Jamshidi, Materials and their characterization, properties, and potential applications in different fields, 2020, DOI: 10.5772/intechopen.92707.

37. J. L. Gunjakar et al., Self-assembly of layered double hydroxide 2D nanoplates with graphene nanosheets: An effective way to improve the photocatalytic activity of 2D nanostructured materials for visible light-induced O_2 generation, *Energy & Environmental Science*, 6, 1008–1017, 2013.

38. Zhang X et al., 2D materials beyond graphene for high-performance energy storage applications, *Advanced Energy Materials*, 6, 1600671, 2016.

39. L. Shi and T. Zhao, Recent advances in inorganic 2D materials and their applications in lithium and sodium batteries, *Journal of Materials Chemistry A*, 5, 3735–3758, 2017.

40. K. Teramura et al., Photocatalytic conversion of CO_2 in water over layered double hydroxides, *AngewandteChemie International Edition*, 51, 8008–8011, 2012.

Chapter 8

Layered Materials: Prospects as Promising Adsorbent for Dye Removal

Ariharaputhiran Anitha[a] and Amalraj John[b]

[a]*V.V. Vanniaperumal College for Women, Virudhunagar, Tamil Nadu, India*
[b]*Instituto de Química de Recursos Naturales, Universidad de Talca, Chile*

anithaa@vvvcollege.org, jamalraj@utalca.cl

8.1 Introduction

For the growing and developing of urban areas and industrial sectors, greener technologies are needed to overcome issues due to various kinds of pollution. For the sustainment of life, we need fresh air and clean water but the natural sources are contaminated by various kinds of pollutants in different ways. Clean water is essential for living organisms to sustain life and development. Minimizing waste and avoiding contamination are serious tasks in society [1]. The growing field in diverse product applications influences society to use various synthetic chemicals, plastics, and non-degradable products, which are harmful to human health and destroy the environmental nature.

Layered Materials: Synthesis and Applications in Energy and Environmental Remediation
Edited by Alagarsamy Pandikumar, Kandasamy Jothivenkatachalam, and Subramanian Moscow

ISBN 978-981-5129-17-5 (Hardcover), 978-1-003-50831-1 (eBook)
www.jennystanford.com

Water is polluted by various pollutants from synthetic dyes from textiles, dyeing, painting, and pharmaceutical products. These pollutants continuously damage the environment and cause unimaginable harm to the environment. These hazardous dyes and chemicals that enter the aquatic environment reduce light penetration, thereby destroying the aquatic life cycle [2, 3]. The presence of effluent dyes in drinking water causes severe damage to humans and causes severe problems including renal failure, reproductive system diseases, and central nervous system illnesses. Also, intake of contaminated water leads to mutagenic and carcinogenic diseases.

To know a little about dyes is noteworthy here. The dyes are colored substances used in industries to make attractive coloring products, which, in turn, generate a large quantity of colored wastewater [4]. Those dyes are classified into three types: cationic, anionic, and non-ionic, based on their functional groups present in the chemical structure. The textile industry held the first position for the usage of dyes in fibers through the binding of reactive groups of dyes and the fiber structure or surface shells to make colors. The dyes are normally soluble in water and possess acidic properties. After the dyeing process, it produces a massive amount of unused dyes, which come as waste products during the process.

According to reports, the textile industry is using more than 10000 synthetic dyes and the amount of consumption is about 1000 tons/year. An estimated 10–15% of unreacted dyes are discharged into the water and effluents affect the ecosystem of biological life [5]. Those reasons create the depolarization of dye-contaminated water is the greatest challenge for the textile industry. The poor degradability and decomposition of dyes from the textile industry cause the major sources of water pollution, i.e., it accounts for 20% of global clean water pollution [6].

The degradation/decomposition and the method of removing unfixing dyes are done by conventional wastewater treatment methods, which include membrane filtration, adsorption, coagulation, flocculation, biological treatment, ion exchange process, and electrochemical technique. The above mentioned methods have the drawbacks of not completely removing the dye molecule from the water. Some of them are not suitable for the industrial process due to high cost and well-equipped instruments; for example, membrane filtration and ion exchange processes. Based on the quality of removal and the cost, the adsorption method is the simplest one compared with the other treatment methods. Due to this reason, it performs in simple operation steps, the adsorbent can be recycled and reused, and showed high efficiency in the removal of dyes. It follows the physiochemical treatment that follows the passive separation of adsorbate from an aqueous phase onto a solid phase. Most of the adsorbent materials used are from natural sources, such as carbon materials and zeolites [7]. Recent research notifies that layered materials act as excellent adsorbents for various dyes.

It is worthwhile to know about layered materials in detail here. Layered materials are solids formed by the stacking of layers or sheets of atoms linked together through strong chemical bonds by exerting weak intermolecular forces of attraction. Layered materials are of different types viz. graphite, boron nitride, dichalcogenides, layered double hydroxides, MoS_2, MXenes, clay, and metal-organic framework (MOF). All these layered materials are capable of adsorbing synthetic dyes from aqueous solutions to a greater extent at a suitable temperature and pH. Various researchers preferred layered materials for dye adsorption for a wide spectrum of dyes over other adsorbents due to their high dye removal efficiency, owing to their high surface area and the possibility of adsorption at various sites, simple operation, richer reusability, and economic sustainability.

This chapter focuses on the use of layered materials for the effective removal of reactive dyes and the required optimum conditions. Because layered materials possess more active surface area than normal catalyst materials, they efficiently remove unfixing molecules and show higher efficiency than any other adsorbent. The effect of the removal and recycling process makes the layered materials suitable nanomaterials for the removal of dyes in industries [6].

8.2 Layered Double Hydroxides (LDH)

Hydrotalcite-double layered hydroxides also known as anionic clay. It has a huge interlayer distance with positively charged layers with interchangeable anions, which has a general formula of $[M_{1-x}^{2+}M_x^{3+}(OH)_2]^{x+}$ $[A_{x/n}^{n-}]^{x-}\cdot mH_2O$. It has the presence of divalent and partially trivalent metal cations responsible for the creation of positively charged layers. The anions are intercalated into the interlayer distance of the metal layer and neutralize the charges on the layered materials. Nowadays, researchers focus on LDH as an effective adsorbent for the removal of dyes in wastewater.

Shan et al. [3] studied Mg–Al-CO_3-LDH as the suitable adsorbent for the degradation of red dyes (reactive red, Congo red, and acid red 1). The LDH was prepared using the co-precipitation method with a mixture ratio of 2:1 for Mg/Al^{3+}. The Mg–Al-CO_3-LDH effectively removes red dyes, and it follows the pseudo-second-order kinetics and the isotherm corresponds to the Langmuir isotherm model. The optimum dosage for the LDH was 10 g and it requires 60 minutes for the degradation product. The mechanism followed for the degradation of dye products, and the electrostatic charge interaction between the LDH and the Congo red dye was established. The adsorption capacity of Mg–Al-LDH for RR, CR, and AR1 was 59.49, 37.16, and 108.0 mg g^{-1}, respectively. The Mg–Al-LDH showed a maximum efficiency of 90%, on the removal of the dye.

Different LDH materials used to remove various dyes, the required optimum conditions, and their efficiencies are tabulated in Table 8.1.

8.3 Graphite

Graphite is a layered material formed by stacking of graphene sheets. Due to its high surface area, surface oxygen functional groups, and the expanded layered structure of graphene that accounts for the excellent adsorbing property, graphite and graphene oxides excel as promising adsorbents for dye removal.

Table 8.1 List of LDH materials used as adsorbents, optimum conditions, and their efficiencies

S. No.	Material	Dye	pH	Temperature (°C)	Adsorption time (min)	Efficiency (mg/g)	Ref.
1	Zn_2Al LDH	Orange II	6	100	15	126.0	[8]
2	C/NiFe-LDH	Methyl orange	7	30	100	323.6	[9]
		Congo red	7	30	100	448.4	
3	$MgAlCO_3$	Amaranth	4	Room	180	200.0	[10]
		Diamine green B	8	Room	180	305.0	
4	$MgAlNO_3$	Amaranth	4	Room	360	800.0	
		Diamine green B	7	Room	360	1089.0	
5	Mg–Al LDH	Reactive red	5	40	120	346.8	[11]
		Reactive blue	5	40	150	389.6	
6	Mg/FeLDH	Congo red	3	Room	180	9127.0	[12]
7	MgAl/BioChar composite	Methylene blue	10	100	100	91.4	[13]
		Rhodamine B	10	40	60	69.2	

Table 8.1 *(Contiued)*

S. No.	Material	Dye	pH	Temperature (°C)	Adsorption time (min)	Efficiency (mg/g)	Ref.
8	Zn–Fe LDH	Methyl orange	7	35	240	230.6	[14]
		Methylene blue	9	35	240	133.2	
		Malachite green	6	35	240	57.3	
9	CoAl LDH	Methyl orange	NA	25	10	816.0	[15]
		Coomassie brilliant blue	NA	25	10	654.8	
		Congo red	NA	25	1	178.3	
		Eosin B	NA	25	1	95.1	
10	Ni/Al LDH	Congo red	NA	Room	120	116.2	[16]
11	MgAl LDH	Congo red	8	50	120	285.71	[17]

NA: Not available.

Jingjie et al. [18] utilized graphene oxide as an adsorbent to remove the Rhodamine B (RhB) dye. The effect of adsorbent dosage, pH, contact time, and initial dye concentration was studied at room temperature. The article reported that the optimal conditions are an initial RhB concentration of 10 mg L^{-1}, an adsorbent amount of 0.24 g, a sonication time of 23 min, and pH 7. The maximum adsorption efficiency was found to be 107.230 mg g^{-1}. The adsorption equilibrium was studied with Langmuir, Freundlich, Langmuir-Freundlich, Redlich-Peterson, Toth, and Khan isotherm models. The Langmuir-Freundlich isotherm model fits well with equilibrium data. The adsorbent has the reusability of up to four cycles with a significant change in efficiency.

Different graphite-based materials used to remove various dyes, the materials used, the required optimum conditions, and their efficiencies are tabulated in Table 8.2.

8.4 Boron Nitride (BN)

In boron nitride, B is bonded with N by covalent bonds and the atomic layers of boron nitride are stacked together by weak intermolecular forces. It possesses unique properties viz. electrical insulation, high oxidation resistance, and outstanding chemical stability [29, 30]. In recent years, BN products with special morphologies like nanosheets, nanocarpets, and nanorods have been widely used as adsorbents, which are discussed below.

Another similar analog of graphene–boron nitride is used as an adsorbent for the removal of carcinogenic dyes-MB and methyl orange. For the enhancement of activity polymer nanocomposite–boron nitride/poly aniline was applied for the degradation studies. The nanocomposite was prepared by the in situ polymerization method and the formation of a polymer composite was confirmed by XRD, TGA, and TEM images. The application of the nanocomposite was suited for the degradation of dyes and it showed that the efficiency was 93% and 95% for both the dyes compared with the individual h-BN and the PANI polymer. The reusability of the nanocomposite showed that the efficiency was maintained at about 74.2% and 78% for the methylene blue and methyl orange dyes after the continuous usage of five cycles. It indicated that the h-BN/PANI nanocomposite acts as a suitable adsorbent for the degradation of both dyes [31].

Table 8.2 List of graphite-based materials used as adsorbents, optimum conditions, and their efficiencies

S. no.	Material	Dye	pH	Temperature (°C)	Adsorption time (min)	Efficiency (mg/g)	Ref.
1	Graphite oxide	Methylene blue	4.0	25	40	351.0	[19]
		Malachite green	4.0	25	40	248.0	
2	Graphite nanoplatelets	Indosol SFGL	3.5	25	120	84.5	[20]
		Drimaren HFRL blue	3.5	25	120	104.3	
3	Graphite oxide	Methylene blue	12.0	20	5	751.9	[21]
4	Red mud-activated graphite composites	Methylene blue	6.0	Room	120	89.2	[22]
5	Multi-walled carbon nanotube-graphite composite	Basic blue 41	6.8	45	87	115.0	[23]
6	Graphite oxide	Methylene blue	NA	Room	150	199.2	[24]
		Crystal violet	NA	Room	80	195.4	
		Rhodamine B	NA	Room	240	154.8	

S. no.	Material	Dye	pH	Temperature (°C)	Adsorption time (min)	Efficiency (mg/g)	Ref.
7	Graphene nanosheets	Direct red 81	3.7	25	360	77.8	[25]
		Indosol SFGL direct blue	3.5	25	360	97.6	
8	Graphene oxide	Direct red 81	7.0	25	360	173.7	
		Indosol SFGL direct blue	7.5	25	360	129.5	
9	Mesoporous graphitic carbon	Safranin dye	6.0	25	365	20.6	[26]
10	Graphene/polyaniline/ Fe_3O_4 nanocomposites	Congo red, even in the presence of brilliant green	4.0	25	120	248.7	[27]
11	Graphene oxide	Methylene blue	6.0	25	50	403.3	[28]
		Rhodamine B	6.0	25	50	686.6	

NA: Not available.

Table 8.3 List of boron nitride-based materials used as adsorbents, optimum conditions, and their efficiencies

S. no.	Material	Dye	pH	Temperature (°C)	Adsorption time (min)	Efficiency (mg/g)	Ref.
1	Boron nitride fibers	Methylene blue	8.0	30	10	392.2	[32]
2	Hexagonal boron nitride	Methylene blue	NA	20	20	230.0	[33]
3	Hexagonal boron nitride	Metanil yellow	4.0	25	180	211.0	[34]
		Victoria blue B	5.0	25	840	895.2	
4	Porous boron nitride	Rhodamine B	NA	25	240	245.0	[35]
5	Boron nitride–carbon nitride composite	Neutral red	6.9	25	480	1350.1	[36]
		Malachite green	7.0	25	480	1040.6	
6	Porous BN nanosheets	Congo red	NA	25	180	782.0	[37]
		Methylene blue	NA	25	180	313.0	
		Basic yellow		25	180	556.0	
7	Cobalt-doped boron nitride nanosheets	Methylene blue	8.5	25	13	340.0	[38]
8	Flake boron nitride	Rhodamine B	3.2	Room	1	125.0	[39]
		Congo red	3.2	Room	270	70.9	
9	Boron nitride nanosheets	Methylene blue	2.0	Room	150	333.3	[40]
		Rhodamine B	8.0	Room	150	312.5	
10	Hexagonal boron nitride	Methyl orange	2.0	Room	40	120.8	[41]
		Direct red 80	2.0	Room	40	328.2	

NA: Not available.

Different boron nitride-based materials used to remove various dyes, the materials used, the required optimum conditions, and their efficiencies are tabulated in Table 8.3.

8.5 Molybdenum Disulfide (MoS_2)

MoS_2, an inorganic analog to graphene, is a multifunctional material and a two-dimensional layered inorganic compound. This crystalline mineral possesses several extraordinary properties, thereby intended for various functional applications such as an adsorbent, photocatalyst, sensor, and membrane for environmental applications. MoS_2-based materials have profound utility as potential adsorbents for various dye molecules. These materials display excellent dye removal performance, owing to their suitable structure, high surface area, large number of adsorption sites, and hierarchical arrangement of MoS_2 in the microscopic spheres, where dye molecules have very fast accessibility [42].

In 2016, Massey et al. [43] prepared microspheres of MoS_2 nanosheets by one-step facile and hydrothermal route from polyethylene glycol and used the prepared MoS_2 as a regenerative adsorbent to remove various dyes from wastewater. An adsorption kinetic study was performed on methylene blue, which illustrated the pseudo-second-order model. The hierarchical microspheres of MoS_2 nanosheets showed high adsorption capacity of 297, 204, 216, 183, and 146 mg g^{-1} for methylene blue, malachite green, rhodamine 6G, fuchsin acid, and Congo red dyes, respectively. All these dyes are removed more than 80% within 50 minutes. The adsorbents can be regenerated and reused efficiently for five subsequent times.

Different MoS_2-based materials used to remove various dyes, the materials used, the required optimum conditions, and their efficiencies are tabulated in Table 8.4.

Table 8.4 List of MoS_2-based materials used as adsorbents, optimum conditions, and their efficiencies

S. no.	Material	Dye	pH	Temperature (°C)	Adsorption time (min)	Efficiency (mg/g)	Ref.
1	MoS_2 nanopowder	Congo red	3.0	50	180	80.6	[44]
2	1T-MoS_2	Methylene blue	8.0	Room	120	787.0	[45]
3	MoS_2 microsphere	Methylene blue	NA	Room	120	84.5	[46]
4	2D MoS_2 nanosheets	Rhodamine B	3.0	Room	10	62.8	[47]
		Methylene blue	3.0	Room	10	58.4	
5	SnS_2/MoS_2 NPs	Rhodamine B	3.0	20	15	125.0	[48]
		Methylene blue	3.0	20	15	202.0	
		Crystal violet	3.0	20	15	165.0	
		Malachite green	3.0	20	15	175.0	
6	MoS_2/S-doped graphitic carbon nitride	Methylene blue	8.0	30	22	166.0	[49]
7	Porous MoS_2	Rhodamine B	NA	Room	420	163.0	[50]
		Methylene blue	NA	Room	420	499.0	
		Methyl orange	NA	Room	420	125.1	

NA: Not available.

8.6 MXene

MXenes are layered materials of two-dimensional transition metal carbides and carbonitrides, having unique structures, multiple active sites, and super-hydrophilic and extraordinary characteristics [51]. By the conventional method, MXenes are produced by immersing MAX phases in concentrated hydrofluoric acid [52]. Among these, the specific formula of MAX is $M_{n+1}AX_n$, where M represents an early transition metal, A is IIIA or IVA group element, and X is C or N, and the value of n is 1, 2, or 3 [53]. On selectively etching $M_{n+1}AX_n$ with hydrofluoric acid, many terminal groups are introduced on its surface and the material obtained has the formula $M_{n+1}X_nT_x$, and T_x is surface terminating groups (T_x = O, OH, F), where x is the number of terminating groups. Due to their outstanding properties, the MXenes and their functional composites have been widely used for different applications such as energy storage and conversion, environmental adsorption, catalysis, and even biomedical fields [54, 55]. In environmental adsorption applications, the MXenes are efficient adsorbents for the removal of environmental pollutants viz. heavy metal ions, organic dyes, and other pollutants [56, 57].

In 2021, Tran et al. [58] used $Ti_3C_2T_x$ MXene as a strong adsorbent for the removal of methylene blue (MB) from wastewater. MXene was prepared by Al-selective etching of Ti_3AlC_2 MAX phase using HF. At a standard condition, the MXene turned out to be terminated mostly with F functional groups. Remarkably, the F-terminated MXene adsorbed approximately 92% of MB in 20 μM of MB aqueous solution within 5 min. It was inferred that the surface termination groups could be organized simply by adjusting cleaning cycles, and the MB removal efficiency had a strong dependence on the pH of MXene suspension. Additionally, the MXene exhibited relatively good recyclability, enabling its repeated use. The adsorption capability of the MXene proves its utility for the remediation of polluted water.

Different MXene-based materials used to remove various dyes, the materials used, the required optimum conditions, and their efficiencies are tabulated in Table 8.5.

Table 8.5 List of MXene-based materials used as adsorbents, optimum conditions, and their efficiencies

S. no.	Material	Dye	pH	Temperature (°C)	Adsorption time (min)	Efficiency (mg/g)	Ref.
1	Nb_2CT_x	Methylene blue	7.0	Room	10	496.0	[59]
		Methyl orange	7.0	Room	10	493.0	
2	$Ti_3C_2T_x$	Methylene blue	NA	20	480	38.9	[60]
3	$Ti_3C_2T_x$ LiOH: $Ti_3C_2T_x$ NaOH: $Ti_3C_2T_x$ KOH: $Ti_3C_2T_x$	Methylene blue	NA	20	120	99.9 121.0 189.0 77.0	[61]
4	Ti_3C_2	MG	6.0	50	150	211.0	[62]
5	$Ti_3C_2T_x/Fe_3O_4$	Methylene blue	5.0	Room	60	153.0	[63]
		Rhodamine B	6.0	Room	60	86.0	
6	MXene	Methylene blue	NA	45	60	86.8	[64]
	MXene-COOH			45	60	70.9	
	MXene-COOH@(PEI/PAA)10			45	60	81.9	
7	$Ti_3C_2–SO_3H$	Methylene blue	11.0	25	60	3739.6	[65]

NA: Not available.

8.7 Clay

Clay minerals are omnipresent in the landscape. Most clay minerals comprise tetrahedral and octahedral sheets, which are classified into two groups based on structural arrangements as 1:1 or 2:1 layers. For example, montmorillonites belong to a class of 2:1 clay with one Al^{3+} octahedral sheet sandwiched between two Si^{4+} tetrahedral sheets. The isomorphous substitution of Al^{3+} for Si^{4+} in the tetrahedral layer and the substitution of Mg^{2+} or Zn^{2+} for Al^{3+} in the octahedral layer leads to a permanent negative charge on the surface of the clay mineral. The layers are held together by van der Waals forces (Lee and Tiwari, 2012). Due to these weak forces and the charge deficits in the structure, water can easily penetrate the layers and some inorganic cations (Na^{+}, Ca^{2+} ions, etc.) balance the negative charge. With lower charge density types of surfaces, clay minerals can play an important role in the attraction of neutral organic contaminants, thus clay minerals such as montmorillonite, kaolinite, illite, and saponite are used as adsorbents. Due to the hydrophilicity of their interlayer surfaces, the adsorption capacity of natural clay minerals toward hydrophobic organic compounds is limited. So, surfactant-modified clay minerals with improved hydrophobicity on their surfaces interact strongly with organic matter.

El Mouzdahir [66] used raw clay minerals to adsorb methylene blue dye from aqueous solutions. The adsorbed quantity of MB on raw clay was 350 mg g^{-1}. So, the raw clay mineral was thermally treated at 300 °C for 2 h and activated with nitric acid under reflux conditions to improve its adsorption capacity. Maximum adsorption of 500 mg g^{-1} of MB at equilibrium is achieved by activated clay at room temperature and atmospheric pressure.

Different clay materials used to remove various dyes, the materials used, the required optimum conditions, and their efficiencies are tabulated in Table 8.6.

Table 8.6 List of clay materials used as adsorbents, optimum conditions, and their efficiencies

S. no.	Material	Dye	pH	Temperature (°C)	Adsorption time (min)	Efficiency (mg/g)	Ref.
1	Fe_3O_4/activated montmorillonite	Methylene blue	7.3	30	25	47.9	[67]
2	Montmorillonite	Basic yellow 2	5.8	60	20	434.1	[68]
3	Moroccan illitic clay	Methylene blue	5.0	40	5	13698	[69]
4	Natural illitic clay	Methylene blue	4.0	40	50	24.8	[70]
5	Natural untreated clay	Basic yellow 2	5.1	25	120	833.3	[71]
6	Activated clay	Methyl orange	7.0	20	30	15.8	[72]
7	Unmodified bentonite	Brilliant blue	6.5	30	2880	6.1	[73]
	Iron-Modified bentonite	Brilliant blue	9.0	30	1440	14.2	[74]
8	Natural Turkish clay	Basic red	10.0	20	30	530.6	[75]
9	Natural palygorskite clay	Methylene blue	7.0	30	720	48.3	[76]
	Heat-treated palygorskite clay	Methylene blue	7.0	30	720	78.1	
10	Natural clay	Acid red 88	2.0	45	15	1133.1	[77]

NA: Not available.

8.8 Metal Organic Framework (MOF)

In recent years, MOF materials have gained worldwide interest for gas separation, membrane permeability, electrochemical catalysis, photocatalysis, and adsorption, due to their large surface area and mesoporous structure [77]. Moreover, it excels in the adsorption of low-molecular-weight compounds from water. Studies have reported that MOF showed excellent selective adsorption capacities toward different kinds of organics by surface modification [78].

Li et al. [79] employed Cu-BTC, a kind of MOF to selectively adsorb MB from dye mixed wastewater. Characterization results revealed that Cu-BTC has an octahedral structure, with a specific surface area of 45.16 m^2 g^{-1} and a pore size of 35–40 nm. The influence of various parameters including the initial solution pH, temperature, ionic strength, initial concentration, and contact time on MB adsorption by Cu-BTC was investigated in detail. The adsorption capacity of Cu-BTC toward MB was optimized at the pH value of 8, with a lower temperature and a higher ionic strength. The maximum adsorption of MB (96%) over Cu-BTC was attributed to the combined effects of hydrophobicity and electrostatic adsorption.

Different MOF materials used to remove various dyes, the materials used, the required optimum conditions, and their efficiencies are tabulated in Table 8.7.

8.9 Conclusion

The organic pollutant removal and the degradation of dye studied by different layered materials were seen. Each of them had the specific property of adsorption and the degradation studies of dyes. The mechanism for the removal process and the efficiency were illustrated in the review.

Though various layered materials excel well as an adsorbent for dye removal, lower regeneration of adsorbents hinders their large-scale practical applications and commercialization.

Table 8.7 List of MOF materials used as adsorbents, optimum conditions, and their efficiencies

S. no.	Material	Dye	pH	Temperature (°C)	Adsorption time (min)	Efficiency (mg/g)	Ref.
1	Iron terephthalate MOF-235	Methyl orange	NA	45	60	501.0	[80]
		Methylene blue	NA	45	60	252.0	
2	Cr-BDC MIL-101	Methyl orange	NA	45	60	140.0	[81]
3	MIL-100(Fe)	Malachite green	4.0	50	220	485.0	[82]
4	MIL-53(Al)-NH_2	Methylene blue	8.0	35	30	208.3	[83]
		Malachite green	8.0	35	30	164.9	
5	MIL-100 (Fe)	Methylene blue	5.0	30	10	645.3	[84]
	MIL-100 (Cr)	Methylene blue	5.0	30	10	736.2	
6	Zirconium MOF (UiO-66)	Alizarin red	2.0	25	36	400.0	[85]
7	UiO-66	Rhodamine B	4.0	50	60	256.4	[86]
8	PCN-222	Methylene blue	8.0	60	45	906.0	[87]
		Methyl orange	8.0	60	45	589.0	
9	TMU-8	Reactive black 5	2.0	50	6	79.3	[88]

NA: Not available.

This can be overcome in the near future by possessing deep knowledge of the materials and the required modification of them with scientific and engineering viewpoints, exploring cost-effective and advanced materials with superior reusability, wide utilization, high efficiency, and long-term stability.

In conclusion, recent developments in science and technology allow many researchers to fabricate different types of layered materials and utilize them for different possible applications of dye degradation and decomposition of organic pollutants. From this chapter, it is clear that the layered materials give viable solutions to the environmental problem of water pollution by organic dyes and other contaminants. The role of layered materials as promising adsorbents for dye removal is highly appreciated. Further research in this field provides a meritorious platform for accelerating its practical applications.

References

1. Starukh, G. (2017). Photocatalytically enhanced cationic dye removal with Zn-Al layered double hydroxides. *Nanoscale Res. Lett.,* 12(1), pp. 391–398.
2. Zhu, S.R. Wu, M.K. Zhao, W.N. Liu, P.F. Yi, F.Y. Li, G.C. Tao, K. and Han, L. (2017). In situ growth of metal–organic framework on BiOBr 2D material with excellent photocatalytic activity for dye degradation. *Cryst. Growth Des.,* 17(5), pp. 2309–2313.
3. Shan, R. Yan, L. Yang, Y. Yang, K. Yu, S. Yu, H. Zhu, B. and Du, B. (2015). Highly efficient removal of three red dyes by adsorption onto Mg–Al-layered double hydroxide. *J. Ind. Eng. Chem.,* 21, pp. 561–568.
4. Guan, S. Zhao, K.F., Qin-Tong, Rao, Q.X. Lin-Cheng, Wei-Song, Zhang, Q.C. Wang, X.L. and Song, W.G. (2021). A review of photocatalytic materials application on nonylphenol degradation. *Environ. Challenges,* 4, pp. 100172–100182.
5. Hao, S. Zhao, X. Cheng, Q. Xing, Y. Ma, W. Wang, X. Zhao, G. and Xu, X. (2020). A mini review of the preparation and photocatalytic properties of two-dimensional materials. *Front. Chem.,* 8, pp. 582146.
6. Royer, B. Cardoso, N. F. Lima, E. C. Macedo, T. R. and Airoldi, C. (2010). A useful organofunctionalized layered silicate for textile dye removal. *J. Hazard. Mater.,* 181(1), pp. 366–374.

7. Ahmed, D.N. Naji, L.A. Faisal, A.A.H. Al-Ansari, N. and Naushad, M. (2020). Waste foundry sand/MgFe-layered double hydroxides composite material for efficient removal of Congo red dye from aqueous solution. *Sci. Rep.,* 10(1), pp. 2042.
8. Andra T. Ioana C. Laura C. Lavinia L. and Gerlinde R. (2020). Adsorption of orange II onto Zn_2Al–layered double hydroxide prepared from zinc ash. *Front. Chem.*, 30, pp. 573535.
9. Hu, H. Wageh, S. Al-Ghamdi, A. A. Yang, S. Tian, Z. Cheng, B. and Ho, W. (2020). NiFe-LDH nanosheet/carbon fiber nanocomposite with enhanced anionic dye adsorption performance. *Appl. Surface Sci.,* 511, pp. 145570.
10. Abdellaoui, K. Pavlovic, I. and Barriga, C. (2019). Nanohybrid layered double hydroxides used to remove several dyes from water. *Chem. Engg.,* 3(2), pp. 41.
11. Boubakri, S. Djebbi, M.A. Bouaziz, Z. Namour, P. Jaffrezic-Renault, N. Amara, A. B. H. and Kalfat, R. (2018). Removal of two anionic reactive textile dyes by adsorption into MgAl-layered double hydroxide in aqueous solutions. *Env. Sci. Pollution Res.,* 25(24), pp. 23817–23832.
12. Ahmed, D.N. Naji, L.A. Faisal, A.A.H. Al-Ansari, N. and Naushad, M. (2020). Waste foundry sand/MgFe-layered double hydroxides composite material for efficient removal of Congo red dye from aqueous solution. *Sci. Rep.,* 10(1), pp. 2042.
13. Ahmed, D.N. Naji, L.A. and Ayad A. H. (2020). Cationic dye removal by magnesium aluminum-biochar composite from aqueous solution, *Int. J. Sci. Technol. Res.*, 9(7), pp. 186–190.
14. Mahmoud, R.K. Taha, M. and Zaher, A. (2021). Understanding the physicochemical properties of Zn–Fe LDH nanostructure as sorbent material for removing of anionic and cationic dyes mixture. *Sci. Rep.,* 11, pp. 21365.
15. Huang, P., Liu, J., Wei, F., Zhu, Y., Wang, X., Cao, C. and Song, W. (2017). Size-selective adsorption of anionic dyes induced by the layer space in layered double hydroxide hollow microspheres. *Mater. Chem. Front.,* 1(8), pp. 1550–1555.
16. Patimah Mega Syah Bahar Nur Siregar, Neza Rahayu Palapa, Alfan Wijaya, Erni Salasia Fitri, and Aldes Lesbani (2021). Structural stability of Ni/Al layered double hydroxide supported on graphite and biochar toward adsorption of Congo red. *Indones. J. Sci. Technol.,* 6(2), pp. 85–95.

17. Ahmed Zaghloul, Abdeljalil Ait Ichou, Mhamed Abali, Ridouan Benhiti, Amina Soudani, Gabriela Carja, Mohamed Chiban, Mohamed Zerbet, and Fouad Sinan (2021). Removal and comparative adsorption of anionic dye on various MgAl synthetic clay. *Biointerface Res. Appl. Chem.*, 11(6), pp. 14986–14997.

18. Jingjie, Y. Siroos, S. and Saeed, S. (2022). Removal of drug and dye from aqueous solutions by graphene oxide: Adsorption studies and chemometrics methods. *npj Clean Water*, 5, pp. 1–10.

19. Bradder, P. Ling, S. K. Wang, S. and Liu, S. (2011). Dye adsorption on layered graphite oxide. *J. Chem. Eng. Data*, 56(1), pp. 138–141.

20. Carvallho, M.N. da Silva, K.S. Sales, D.C.S. Freire, E.M.P.L. Sobrinho, M.A.M. and Ghislandi, M.G. (2016). Dye removal from textile industrial effluents by adsorption on exfoliated graphite nanoplatelets: kinetic and equilibrium studies. *Water Sci. Technol.*, 73(9), pp. 2189–2198.

21. He, G., Zhang, J., Zhang, Y., Chen, H. and Wang, X. (2013). Fast and efficient removal of cationic dye using graphite oxide, adsorption, and kinetics studies. *J. Dispers. Sci. Technol.*, 34(9), pp. 1223–1229.

22. Linh, H.X. Thu, N.T. Toan, T.Q. Huong, D. T. Giang, B. T. Ha, H. K. P. and Hai, N.T. (2019). Fast and effective route for removing methylene blue from aqueous solution by using red mud-activated graphite composites. *J. Chem.*, pp. 2858170.

23. Ali Shirgholami, M. Mirjalili, M. and Nasirizadeh, N. (2017). Preparation of aMWCNT-graphite composite based on sol gel method for dye removal. *Orient. J. Chem.*, 33(2), pp. 676–685.

24. Jin, Q.Q. Zhu, X.-H. Xing, X.Y. and Ren, T.Z. (2012). Adsorptive removal of cationic dyes from aqueous solutions using graphite oxide. *Adsorp. Sci. Technol.*, 30(5), pp. 437–447.

25. De Assis, L.K. Damasceno, B.S. Carvalho, M.N. Oliveira, E.H.C. and Ghislandi, M. G. (2019). Adsorption capacity comparison between graphene oxide and graphene nanoplatelets for the removal of colored textile dyes from wastewater. *Environ. Technol.*, 41(18), pp. 2360–2371.

26. Shaban, M., Abukhadra, M. R., Mohamed, A. S., Shahien, M. G. and Ibrahim, S. S. (2017). Synthesis of mesoporous graphite functionalized by nitrogen for efficient removal of safranin dye utilizing rice husk ash: Equilibrium studies and response surface optimization. *J. Inorg. Organomet. Polym. Mater.*, 28(1), pp. 279–294.

27. Mu, B. Tang, J. Zhang, L. and Wang, A. (2017). Facile fabrication of superparamagnetic graphene/polyaniline/Fe_3O_4 nanocomposites for fast magnetic separation and efficient removal of dye. *Sci. Rep.,* 7(1), pp. 5347.
28. Mao, B. Sidhureddy, B. Thiruppathi, A.R. Wood, P.C. and Chen, A. (2020). Efficient dye removal and separation based on graphene oxide nanomaterials. *New J. Chem.*, 44, pp. 4519–4528.
29. Singhal, S.K. Srivastava, A.K. Pant, R.P. Halder, S.K. Singh, B.P. and Gupta, AK. (2008). Synthesis of boron nitride nanotubes employing mechanothermal process and its characterization. *J. Mater. Sci.,* 43(15), pp. 5243–5250.
30. Zheng, Z.Y. Cox, M. and Li, B. (2018). Surface modification of hexagonal boron nitride nanomaterials: A review. *J. Mater. Sci.,* 53(1), pp. 66–99.
31. Shahabuddin, S. Khanam, R. Khalid, M. Sarih, N. M. Ching, J. J. Mohamad, S. and Saidur, R. (2018). Synthesis of 2D boron nitride doped polyaniline hybrid nanocomposites for photocatalytic degradation of carcinogenic dyes from aqueous solution. *Arab. J. Chem.,* 11(6), pp. 1000–1016.
32. Li, J. Huang, Y. Liu, Z. Zhang, J. Liu, X. Luo, H. and Tang, C. (2015). Chemical activation of boron nitride fibers for improved cationic dye removal performance. *J. Mater. Chem. A,* 3(15), pp. 8185–8193.
33. Shen, T. Liu, S. Yan, W. and Wang, J. (2019). Highly efficient preparation of hexagonal boron nitride by direct microwave heating for dye removal. *J. Mater. Sci.,* 54(12), pp. 8852–8859.
34. Tarhan, T. (2020). Synthesis and characterization of hexagonal boron nitride used for comparison of removal of anionic and cationic hazardous azo-dye: Kinetics and equilibrium studies. *Turk. J. Chem.*, 44(6), pp. 1471–1482.
35. Marchesini, S. Wang, X. and Petit, C. (2019). Porous boron nitride materials: Influence of structure, chemistry and stability on the adsorption of organics. *Front. Chem.,* 7, pp. 160.
36. Guo, Y. Wang, R. Wang, P. Rao, L. and Wang, C. (2019). Developing novel layered boron nitride-carbon nitride composite with high efficiency and selectivity to remove protonated dyes from water. *ACS Sustain. Chem. Eng.,* 7(6), pp. 5727–5741.
37. Lei, W. Portehault, D. Liu, D. Qin, S. and Chen, Y. (2013). Porous boron nitride nanosheets for effective water cleaning. *Nat. Commun.,* 4, pp. 1777.

38. Hassan, J. Ikram, M. Ul-Hamid, A. Imran, M. Aqeel, M. and Ali, S. (2020). Application of chemically exfoliated boron nitride nanosheets doped with Co to remove organic pollutants rapidly from textile water. *Nanoscale Res. Lett.,* 15(1), pp. 75.

39. Qu, J. Li, Q. Luo, C. Cheng, J. and Hou, X. (2018). Characterization of flake boron nitride prepared from the low temperature combustion synthesized precursor and its application for dye adsorption. *Coatings,* 8(6), pp. 214.

40. Bangari, R.S. and Sinha, N. (2020). Synthesis of high surface area boron nitride nanosheets for the removal of methylene blue and Rhodamine B dyes from contaminated water. *J. Nanosci. Nanotechnol.,* 20(10), pp. 6222–6234.

41. Das, R. Solis-Fernandez, P. Breite, D. Prager, A. Lotnyk, A. Schulze, A. and Ago, H. (2020). High flux and adsorption based non-functionalized hexagonal boron nitride lamellar membrane for ultrafast water purification. *Chem. Eng. J.,* pp. 127721.

42. Liu, C. Wang, Q. Jia F, Song S. (2019). Adsorption of heavy metals on molybdenum disulfide in water: A critical review. *J. Mol. Liq.,* 292, pp. 111390.

43. Massey, A.T. Gusain, R. Kumari, S. and Khatri, O.P. (2016). Hierarchical microspheres of MoS_2 nanosheets: Efficient and regenerative adsorbent for removal of water-soluble dyes. *Ind. Eng. Chem. Res.,* 55(26), pp. 7124–7131.

44. Alarifi, I. M. Al-Ghamdi, Y.O. Darwesh, R., Ansari, M. O. and Uddin, M. K. (2021). Properties and application of MoS_2 nanopowder: Characterization, Congo red dye adsorption, and optimization. *J. Mater. Res. Technol.,* 13, pp. 1169–1180.

45. Omar, A. M., Metwalli, O. I., Saber, M. R., Khabiri, G., Ali, M. E. M., Hassen, A., and Khalil, A. S. G. (2019). Revealing the role of the 1T phase on the adsorption of organic dyes on MoS_2 nanosheets. *RSC Adv.,* 9(49), pp. 28345–28356.

46. Panda, J. and Tudu, B. (2018). Hydrothermally synthesized flower like MoS_2 microsphere: A highly efficient adsorbent for methylene blue dye removal. *AIP Conf. Proc.,* 1953, pp. 030127.

47. Han, S. Liu, K. Hu, L., Teng, F. Yu, P. and Zhu, Y. (2017). Superior adsorption and regenerable dye adsorbent based on flower-like molybdenum disulfide nanostructure. *Sci. Rep.,* 7, pp. 43599.

48. Cui, B., Wang, Y., Zhang, F., Xiao, X., Su, Z., Dai, X. and Huang, S. (2019). Hydrothermal synthesis of SnS_2/MoS_2 Nanospheres for

enhanced adsorption capacity of organic dyes. *Mater. Res. Express,* 7, pp. 015016.

49. Chegeni, M. Mehri, M. and Hosseini, M. (2020). The MoS_2/S-doped graphitic carbon nitride: Synthesis and application as a composite for removing organic pollutant. *J Water Environ Nanotechnol.,* 5(4), pp. 331–441.

50. Li, H. Xie, F. Li, W. Fahlman, B. D. Chen, M. and Li, W. (2016). Preparation and adsorption capacity of porous MoS_2 nanosheets. *RSC Adv.,* 6(107), pp. 105222–105230.

51. Lei, J.C. Zhang, X. and Zhou, Z. (2015). Recent advances in MXene: Preparation, properties, and applications. *Front. Physiol.,* 10, pp. 276–286.

52. Naguib, M. Mashtalir, O. Carle, J. Presser, V. Lu, J. Hultman, L. Gogotsi, Y. and Barsoum, M.W. (2012). Two-dimensional transition metal carbides. *ACS Nano,* 6, pp. 1322–1331.

53. Ng, V.M.H. Huang, H. Zhou, K. Lee, P.S. Que, W. Xu, J.Z. and Kong, L.B. (2017). Recent progress in layered transition metal carbides and/or nitrides (MXenes) and their composites: synthesis and applications. *J. Mater. Chem. A,* 5, pp. 3039–3068.

54. Xue, Q. Zhang, H. Zhu, M. Pei, Z. Li, H. Wang, Z. Huang, Y. Huang, Y. Deng, Q. and Zhou, J. (2017). Photoluminescent Ti_3C_2 MXene quantum dots for multicolor cellular imaging. *Adv. Mater.,* 29, pp. 1604847.

55. Wu, X. Wang, Z. Yu, M. Xiu, L. and Qiu, J. (2017). Stabilizing the MXenes by carbon nanoplating for developing hierarchical nanohybrids with efficient lithium storage and hydrogen evolution capability. *Adv. Mater.,* 29, pp. 1607017.

56. Zhang, Y.J. Zhou, Z.J. Lan, J.H. Ge, C.C. Chai, Z.F. Zhang, P. and Shi, W.Q. (2017). Theoretical insights into the uranyl adsorption behavior on vanadium carbide MXene. *Appl. Surf. Sci.,* 426, pp. 572–578.

57. Wang, L. Song, H. Yuan, L. Li, Z. Zhang, Y. Gibson, J.K. Zheng, L. Chai, Z. and Shi, W. (2018). Efficient U (VI) reduction and sequestration by Ti_2CT_x MXene. *Environ. Sci. Technol.,* 52, pp. 10748–10756.

58. My Tran, N. Thanh Hoai Ta, Q. Sreedhar, A. and Noh, J.S. (2021). $Ti_3C_2T_x$ MXene playing as a strong methylene blue adsorbent in wastewater. *Appl. Surf. Sci.,* 537, pp. 148006.

59. Yan, Y. Han, H. Dai, Y. Zhu, H. Liu, W. Tang, X. Gan, W and Li, H. (2021). Nb_2CT_x MXene nanosheets for dye adsorption. *ACS Appl. Nano Mater.* 4(11), pp. 11763–11769.

60. Mashtalir, O. Cook, K.M. Mochalin, V.N. Crowe, M. Barsoum, M.W. and Gogotsi, Y. (2014). Dye adsorption and decomposition on two-dimensional titanium carbide in aqueous media. *J. Mater. Chem. A*, 2 (35), pp. 14334–14338.

61. Wei, Z. Peigen, Z. Wubian, T. Xia, Q. Yamei, Z. and Zheng Ming, S. (2018). Alkali treated $Ti_3C_2T_x$ MXenes and their dye adsorption performance. *Mater. Chem. Phys.,* 206, pp. 270–276.

62. Soha, M.A. Mohamed, A.S. and Ahad, M. Aldawsari, M. (2022). Removal of malachite green dye from water using MXene (Ti_3C_2) nanosheets. *Sustainability,* 14(10), pp. 5996.

63. Rethinasabapathy, M. Bhaskaran, G. Park, B., Shin, J.Y. Kim, W.S. Ryu, J. and Huh, Y.S. (2022). Iron oxide (Fe_3O_4)-laden titanium carbide ($Ti_3C_2T_x$) MXene stacks for the efficient sequestration of cationic dyes from aqueous solution. *Chemosphere,* 286, pp. 131679.

64. Li, K. Zou, G. Jiao, T. Xing, R. Zhang, L. Zhou, J. and Peng, Q. (2018). Self-assembled MXene-based nanocomposites via layer-by-layer strategy for elevated adsorption capacities. *Colloids Surf. A: Physicochem. Eng. Asp.,* 553, pp. 105–113.

65. Lei, Y. Cui, Y. Huang, Q. Dou, J. Gan, D. Deng, F. and Wei, Y. (2019). Facile preparation of sulfonic groups functionalized Mxenes for efficient removal of methylene blue. *Ceram. Int.,* 45(14), pp. 17653–17661.

66. El Mouzdahir, Y. Elmchaouri, A. Mahboub, R, Gil, A. and Korili, S.A. (2010). Equilibrium modeling for the adsorption of methylene blue from aqueous solutions on activated clay minerals. *Desalination,* 250(1), pp. 335–338.

67. Chang, J. Ma, J. Ma, Q. Zhang, D. Qiao, N. Hu, M. and Ma, H. (2016). Adsorption of methylene blue onto Fe_3O_4/activated montmorillonite nanocomposite. *Appl. Clay Sci.,* pp. 119, 132–140.

68. Sozudogru, O. Fil, B.A. Boncukcuoglu, R. Aladag, E. and Kul, S. (2015). Adsorptive removal of cationic (BY2) dye from aqueous solutions onto Turkish clay: Isotherm, kinetic, and thermodynamic analysis. *Part. Sci. Technol.,* 34(1), pp. 103–111.

69. Amrhar, O. Nassali, H. and Elyoubi, M.S. (2015). Adsorption of a cationic dye, methylene blue, onto moroccan illitic clay. *J. Mater. Environ. Sci.,* 6 (11), pp. 3054–3065.

70. Ozdes, D. Duran, C. Senturk, H. B. Avan, H. and Bicer, B. (2013). Kinetics, thermodynamics, and equilibrium evaluation of adsorptive removal

of methylene blue onto natural illitic clay mineral. *Desalination Water Treat.*, 52(1–3), pp. 208–218.

71. Oztürk, A. and Malkoc, E. (2014). Adsorptive potential of cationic Basic Yellow 2 (BY2) dye onto natural untreated clay (NUC) from aqueous phase: Mass transfer analysis, kinetic and equilibrium profile. *Appl. Surf. Sci.*, 299, pp. 105–115.
72. Ma, Q., Shen, F., Lu, X., Bao, W. and Ma, H. (2013). Studies on the adsorption behavior of methyl orange from dye wastewater onto activated clay. *Desalination Water Treat.*, 51(19–21), pp. 3700–3709.
73. Hernandez-Hernandez, K.A. Solache-Rios, M. and Diaz-Nava, M. C. (2013). Removal of Brilliant Blue FCF from aqueous solutions using an unmodified and iron-modified bentonite and the thermodynamic parameters of the process, *Water Air Soil Pollut.*, 224(5), pp. 1562.
74. Fil, B.A. Karcioglu Karakas, Z. Boncukcuoglu, R. and Yilmaz, A.E. (2013). Removal of cationic dye (basic red 18) from aqueous solution using natural Turkish clay, *Global Nest J.*, 15, pp. 529–541.
75. Chen, H., Zhao, J., Zhong, A., & Jin, Y. (2011). Removal capacity and adsorption mechanism of heat-treated palygorskite clay for methylene blue. *Chem. Eng. J.*, 174(1), pp. 143–150.
76. Akar, S.T. and Uysal, R. (2010). Untreated clay with high adsorption capacity for effective removal of C.I. Acid Red 88 from aqueous solutions: Batch and dynamic flow mode studies. *Chem. Eng. J.*, 162(2), pp. 591–598.
77. Li, J.R. Sculley, J. and Zhou, H.C. (2012). Metal-organic frameworks for separations. *Chem. Rev.*, 112, pp. 869–932.
78. Hasan, Z. and Jhung, S.H. (2015). Removal of hazardous organics from water using metal-organic frameworks (MOFs): Plausible mechanisms for selective adsorptions. *J. Hazard. Mater.*, 283, pp. 329–339.
79. Li, Y. Gao, C. Jiao, J. Cui, J. Li, Z. and Song, Q. (2021). Selective adsorption of metal–organic framework toward methylene blue: Behavior and mechanism. *ACS Omega*, 6 (49), pp. 33961–33968.
80. Haque, E. Jun, J. W. and Jhung, S.H. (2011). Adsorptive removal of methyl orange and methylene blue from aqueous solution with a metal-organic framework material, iron terephthalate (MOF-235). *J. Hazard. Mater.*, 185(1), pp. 507–511.
81. Haque, E. Lee, J. E. Jang, I. T. Hwang, Y. K. Chang, J.S. Jegal, J. and Jhung, S.H. (2010). Adsorptive removal of methyl orange

from aqueous solution with metal-organic frameworks, porous chromium-benzenedicarboxylates. *J. Hazard. Mater.,* 181(1–3), pp. 535–542.

82. Huo, S.H. and Yan, X.P. (2012). Metal–organic framework MIL-100(Fe) for the adsorption of malachite green from aqueous solution. *J. Mater. Chem. A,* 22(15), pp. 7449.

83. Li, C. Xiong, Z. Zhang, J. and Wu, C. (2015). The strengthening role of the amino group in metal–organic framework MIL-53 (Al) for methylene blue and malachite green dye adsorption. *J. Chem. Eng. Data,* 60(11), pp. 3414–3422.

84. Tong, M. Liu, D. Yang, Q. Devautour-Vinot, S. Maurin, G. and Zhong, C. (2013). Influence of framework metal ions on the dye capture behavior of MIL-100 (Fe, Cr) MOF type solids. *J. Mater. Chem. A,* 1(30), pp. 8534.

85. Embaby, M.S. Elwany, S.D. Setyaningsih, W. and Saber, M.R. (2018). The adsorptive properties of UiO-66 towards organic dyes: A record adsorption capacity for the anionic dye Alizarin Red S. *Chin. J. Chem. Eng.,* 26(4), pp. 731–739.

86. He, Q. Chen, Q. Lü, M. and Liu, X. (2014). Adsorption Behavior of Rhodamine B on UiO-66. *Chin. J. Chem. Eng.,* 22(11–12), pp. 1285–1290.

87. Li, H. Cao, X. Zhang, C. Yu, Q. Zhao, Z. Niu, X. and Li, Z. (2017). Enhanced adsorptive removal of anionic and cationic dyes from single or mixed dye solutions using MOF PCN-222. *RSC Adv.,* 7(27), pp. 16273–16281.

88. Hazrati, M. and Safari, M. (2020). Cadmium- based metal–organic framework for removal of dye from aqueous solution. *Environ. Prog. Sustain. Energy,* 39(5), pp. 13411.

Index

Printed in the USA
CPSIA information can be obtained
at www.ICGtesting.com
LVHW071246140124
768959LV00005B/210

9 783031 482618